DATE DUE

DE 18 '96		
MR 31		
NV 20 99		

DEMCO 38-296

Frontispiece. *Drosophila melanogaster* (J.A.L. Cooke/Oxford Scientific Films Ltd).

PETER A. LAWRENCE

MRC LABORATORY OF MOLECULAR BIOLOGY
HILLS ROAD, CAMBRIDGE

The Making of a Fly

THE GENETICS OF ANIMAL DESIGN

OXFORD

BLACKWELL SCIENTIFIC PUBLICATIONS

LONDON EDINBURGH BOSTON

MELBOURNE PARIS BERLIN VIENNA

© 1992 by
Blackwell Scientific Publications
Editorial Offices:
Osney Mead, Oxford OX2 0EL
25 John Street, London WC1N 2BL
23 Ainslie Place, Edinburgh EH3 6AJ
238 Main Street, Cambridge
    Massachusetts 02142, USA
54 University Street, Carlton
    Victoria 3053, Australia

Other Editorial Offices:
Librairie Arnette SA
2 rue Casimir-Delavigne
75006 Paris
France

Blackwell Wissenschafts-Verlag
Meinekestrasse 4
D-1000 Berlin 15
Germany

Blackwell MZV
Feldgasse 13
A-1238 Wien
Austria

First published 1992
Reprinted (with corrections) 1992

Set by Setrite Typesetters, Hong Kong
Printed and bound in Great Britain at
The Alden Press, Oxford

DISTRIBUTORS

Marston Book Services Ltd
PO Box 87
Oxford OX2 0DT
(*Orders*: Tel: 0865 791155
        Fax: 0865 791927
        Telex: 837515)

USA
Blackwell Scientific Publications, Inc
238 Main Street
Cambridge, MA 02142
(*Orders*: Tel: 800 759-6102
        617 876-7000)

Canada
Oxford University Press
70 Wynford Drive
Don Mills
Ontario M3C 1J9
(*Orders*: Tel: 416 441-2941)

Australia
Blackwell Scientific Publications
Pty Ltd
54 University Street
Carlton, Victoria 3053
(*Orders*: 03 347-5552)

British Library
Cataloguing in Publication Data

Lawrence, Peter A.
    The making of a fly.
    I. Title
    591.15

    ISBN 0-632-03048-8

Library of Congress
Cataloging in Publication Data

Lawrence, Peter A. (Peter Anthony),
1941-
    The making of a fly: the genetics of
        animal design/by Peter A. Lawrence.
            p.   cm.
    Includes bibliographical references.
    ISBN 0-632-03048-8
    1. Drosophila melanogaster —
Genetics. 2. Embryology — Insects.
I. Title.
QL537.D76L39   1992
595.77'4 — dc20

To Birgitta and the Royal Shakespeare Company who have kept me almost sane. To my mother who, as well as so many other things, helped feed my caterpillars.

'For more than 25 years I have looked at the little fly *Drosophila* and each time I am delighted anew. When I see it under moderate magnification of a binocular microscope I marvel at the clear-cut form of the head with giant red eyes, the antennae, and elaborate mouth parts; at the arch of the sturdy thorax bearing a pair of beautifully iridescent, transparent wings and three pairs of legs; at the design of the simple abdomen composed of a series of ringlike segments. A shining, waxed armor of chitin covers the whole body of the insect. In some regions this armor is bare; in other regions there arise short or long outgrowths, strong and wide at the base and gently tapering to a fine point. These are the bristles. Narrow grooves, as in fluted columns with a slightly baroque twist, extend along their lengths.' [1]

'To live and not to know why the cranes fly, why children are born, why the stars are in the sky. Either you know why you're alive or it's all nonsense, it's all dust in the wind.' [2]

Contents

Preface

There are several thousand people whose working lives centre around the little fruitfly *Drosophila melanogaster*. In recent years the emphasis of their studies has shifted from inheritance to development. In the hands of a small number of particularly imaginative scientists, traditional genetics, experimental embryology and new molecular techniques have been combined to build a picture of developmental mechanisms. This picture is new and exciting — although it is far from complete it represents the beginnings of a real understanding of how one animal is designed and built. This book, which is written for students and other interested persons, rather than for specialists, aims to give a glimpse of that picture.

The sum of useful information on *Drosophila* would not fit into a hundred books of this size; even now the molecular analysis of new genes is being reported at the rate of one a week and rising. Naturally, I have been selective, favouring subjects that seem important to me. Inevitably, few experts, if any, will approve of my selection. Also, there are a good many opinions in this book and I have tried not to obscure them by prevarication. Many will be superseded or destroyed by new experiments and ifs and buts put in now will be powerless to preserve them then.

One problem for the reader, and the writer, is that often in order to understand something, you need to understand other things and these cannot always have been described previously. What I've tried to do is explain things in an epigenetic order so that explanations are built on each other. It should therefore be possible to read the book through from the beginning and acquire understanding piece upon piece. This means that the book may not be easy to dip into, at least for the novice. By keeping it short, I hope the reader will be able to soldier through it without giving up.

This book is concerned with molecular outcomes but not with molecular techniques; information on cloning methods, cDNA libraries, expression vectors, northern blots, *in situ* hybridisation and the like is available in textbooks. However, I hope it can be understood without this technical knowledge. Other methods are described in outline; in order not to break up the storyline they have been placed in boxes near where they are most needed. The same applies to various pieces of explanation and pontification which are also boxed. Where possible, the figures have a constant orientation, anterior is to the left and ventral downwards.

The book is not completely referenced, as the text would have been too broken up; **it should not be quoted as a reference for matters of fact**. The reader can get into the primary and secondary literature by the lists of experimental papers and reviews; these references have been chosen to give convenient and timely access to the literature and not to acknowledge particular authors. The primary sources of the figures are also given. Generally, I have not tried to apportion credit to the scientists who have made the discoveries — this is becoming increasingly intricate and hazardous and hinders communication. There are, however, six historical short stories which give the background to some of the key discoveries. The figures do not have long legends; they need to be studied in conjunction with the text.

Harvey Shoolman of Blackwell Scientific Publications suggested I write a book modelled in spirit on Mark Ptashne's *A Genetic Switch*. Both of them helped me: Mark at the beginning when I was wondering whether to try it and Harvey was always there to advise and encourage whenever I felt like dropping it. Paul Johnston and Denise Cooper kept the coracle afloat in the lab and Mark Bretscher, as he has for 4 years now, took on responsibilities alone that I should have shared with him. Barbara Cross, drawing on years of practice, decoded my pencilled scripts. Pat Simpson criticised and improved Chapter 7. Gines Morata gave several days to help me revise Chapter 5 and parts of the histories. Gary Struhl read the entire manuscript, gave much valued advice and helped me extirpate some of the nonsense. Adelaide Carpenter scrutinised and honed the whole text with great care. I regret I cannot blame these, my generous friends, for the mistakes that remain.

Rachael Stock edited the text and mollified an anxious author. Edward Wates designed the book, Denys Ovenden drew the original fly pictures and David Gardner spent months patiently transforming my scruffy sketches into figures. Michael Ashburner, Michael Bate, Michael and Susan Berridge, Mariann Bienz, Bonnie, Frankie, Katie and Maggie Bolt, Peter Bryant, Henry Disney, Jim George, Iva Greenwald, Thomas Gutjahr, Ernst Hafen, Anna Haraldson, Jonathan Hodgkin, Herbert Jäckle, the late Ivor Lawrence, Tony Lees, Ruth Lehmann, Michael Levine, Ben Lewin, Ed and Pam Lewis, Bill McGinnis, Juan Modolell, Janni Nüsslein-Volhard, Gerry Rubin, Ernesto Sanchez-Herrero, Klaus Sander, Matthew Scott, Schutt, the late Sydney Smith, Juliet Stevenson, Andrew Tomlinson, Eric Wieschaus, Michael Wilcox, V.B. Wigglesworth and Lewis Wolpert helped in different ways.

Birgitta Haraldson has made a home for me.

Thanks to them all.

P.A.L.
Great Wilbraham
May 1991

Introduction

'At first I could see nothing, the hot air escaping from the chamber causing the candle flame to flicker, but presently, as my eyes grew accustomed to the light, details of the room within emerged slowly from the mist, strange animals, statues and gold — everywhere the glint of gold.' [3]

While unfamiliar things excite curiosity, the everyday miracles of animal and plant development are taken for granted. We enjoy the way seeds transform into flowering plants and caterpillars become butterflies, but generally we accept and do not investigate. There have always been a few scientists who have been interested but until recently they have been regarded as a bit eccentric. Nevertheless, the quest to understand development is fascinating, resembling the search for and discovery of Tutankhamun's tomb, as well as the gradual unpicking of its golden inventory. Hidden in the coded hieroglyphics of the DNA sequence are not only the instructions to make the organism but also an immemorial evolutionary history.

Developmental biology is now in the ascendant and every year hundreds of scientists either become developmental biologists or, at least, so describe themselves. There are several reasons for this. Firstly, there is the ability to purify and sequence the DNA from specific genes, many of which are concerned with building the animal rather than making it work. These advances have transformed embryology; it used to be rather a soft, almost whimsical, science in which lecturers showed slides of beautiful embryos and the horrible malformations they induced by grafting experiments. Now, modern developmental biologists are tough and hard-nosed, they deal in gels, sequences and computers; some do not even look at embryos. Secondly, there is the conviction that one of the great mysteries of life is beginning to crack open and, 'Since things in motion sooner catch the eye than what not stirs' [4], scientists want to be right there as the secrets are discovered. Thirdly, although there has been overall progress in understanding, it is the advances made on convenient organisms that have really impressed, particularly the fruitfly and the nematode.

The advantages of the fruitfly as experimental material are many. Briefly, there are the thousands of man and woman years already invested in studying the genetics and cytogenetics of *Drosophila* which have led to a map of the genes that is far superior to that of any other complex organism. The accuracy of this map has depended on the enormous banded chromosomes that are found only in the salivary glands of flies. Fruitflies breed fast, the life cycle is 10–14 days, and hundreds can be kept in tubes or small milk bottles. They have but three major chromosomes, which only undergo recombination in the female germ cells, not the male. As in any science, progress is not due entirely or even mainly to the chosen systems, important though they are; it is the imagination of a few outstanding people that has been crucial, and some of these are pictured in the short histories at the back of this book.

It has long been an article of faith amongst biologists that understanding gained by studying one system is likely to apply to others and often this has proved to be the case. When I was a student there was a growing subject in insect biochemistry; it was suspected by some that insects might do things very differently — they might not have a Krebs cycle, for instance. Of course it has turned out that insect and mammalian biochemistry are fundamentally similar. Even so, we cannot be certain that this universalist principle will apply at higher levels of organisation. We cannot be certain that the developmental biology of humans will be furthered by studies on fruitflies but we do believe it. More than that, we believe that we can make more rapid progress on vertebrates by working in parallel on smaller animals such as flies and worms. This viewpoint has already earned support as homeoboxes (discovered in flies) have proved a means to isolate mammalian designer genes. Now, it is possible to combine the many advantages of vertebrates, for example in tissue culture systems, with the genetic knowledge of simpler organisms.

One of the first things it would be good to decide, when you approach a vast problem like animal development, is what level of understanding you are looking for. This has usually been an unattainable luxury as **any** kind of information has been sought after. But now there is room for aims and strategy. Do we most want to know how key molecules interact — the minute changes to the structure of a regulatory protein as it binds to DNA? Is it sufficient to list the molecules involved and the outcome of their interactions? Better, perhaps, to see development at the cellular level, to understand the way cells change shape and move as they construct organs? Maybe a geneticist's perspective is more illuminating; when and where are particular genes used and what is the function of their protein products?

In my opinion, the heart of development is the step by step allocation of cells to more and more precisely determined fates. These allocations have to occur in the right part of the embryo. This key process can be broken down into two questions: How are the cells chosen by position? What happens inside a cell when it becomes allocated to a specific fate; that is what are the molecular and genetic changes? A good deal of progress has been made in answering these questions, answers that structure this book.

In what follows, I present an up-to-date but simplified picture of fly development. The field is developing so fast that much of that picture is liable to change; therefore I have tried to avoid ephemera and, where possible, to use well-established material. Workers on flies have gone further than others and gained a deeper understanding of animal design and it is this understanding, as well as the excitement that goes with it, that I wish to convey.

1 The mother and the egg

EARLY DEVELOPMENT is briefly described; the formation of the first layer of cells and its elaboration into more layers by gastrulation, neurogenesis and formation of the gut. Much of the embryo is then divided into parasegments. The mother deposits RNA species representing approximately 80% of all genes into the egg; some of the remainder belong to a special class of controlling genes that would wreak havoc if their products were released freely into the egg.

Which came first, the chicken or the egg? This traditional question is related to the ones that have been at the centre of the oldest and grandest debate in embryology: to what extent does the egg contain a miniature and complete version of the fully developed embryo (the embryo being 'preformed')? Is it more correct to think of the embryo as being progressively elaborated from relatively simple beginnings (that is by 'epigenesis')? It seems to me that although the latter view prevails, the argument rumbles on and still energises much research. Genetics is a method of logic and, when harnessed to modern molecular techniques, has gained in power — it is probably the best way to approach these key questions.

The transition of the generations, the start of the making of a fly, occurs as the mother begins to build the egg. The egg includes storage material, proteinaceous yolk, vesicles of lipid, mitochondria and ribosomes, but it is not known how much spatially organised information it contains. Some kind of positional information is needed to make the pattern of the fly, that is to so organise the cells that they make all the body parts in their proper places; this information must interact with the genetic information carried in the nuclei of the egg and the sperm. It is this faculty, the process of **pattern formation**, that makes a fly a fly or a hippopotamus a hippopotamus.

If the information to build the fly is already deployed in the newly laid egg, then the genes that encode that information must be expressed and utilised while the egg is being constructed, during oogenesis. Such genes will have a 'maternal effect', meaning that the development of the embryo will depend on the genetic status, the **genotype** of the mother. Thus, if a key gene is missing in the mother, the embryo will develop with an abnormal pattern, that is a mutant **phenotype**, regardless of whether or not the father provides that gene in the sperm. If, alternatively, the mother merely provides numerous ingredients (such as myosins and tubulins), the information that specifies the pattern of the embryo (such as transcription factors which turn other genes on and off) must be brought into action **after** fertilisation. In this case, the genotype of the fertilised egg, the zygote, with its contributions from both parents, will be the prime determinant of the embryonic development.

How many gene products needed for pattern formation are placed in the egg by the mother; how many are only made in the zygote? The most direct way to build an answer to this fundamental question is to

---

Box 1.1 Genetic nomenclature

In *Drosophila*, genetic loci are given names according to the mutant phenotype. For example, the famous (the second discovered) mutant was *white[1]* or *w[1]* for short. *white[1]* is recessive, and has therefore a small letter (yes, even at the beginning of a sentence). It is viable and fertile when homozygous. Sometimes, different genes with similar or identical phenotypes are given related names; for example a family of maternal-effect lethals were cleverly named after royal households that became barren and then extinct and were therefore reminiscent of the phenotype of the embryo (*tudor, vasa, valois*).

When the mutant is dominant, it is given a capital letter (such as *Ultrabithorax, Ubx*); but dominance is a capricious feature as it can have various causes. If the mutation merely eliminates functional product from one chromosome, dominance will depend on the dose sensitivity of the phenotype and how hard we look. *Ubx* is only dominant because in heterozygous flies one can see that the haltere is slightly swollen — this gives little hint that the true recessive phenotype is an embryonic lethal. This type of dominance is due to 'haplo insufficiency' — the inability of one wildtype dose of the gene to give a completely normal wildtype phenotype. Dominance can also be caused when the gene is so altered that it either produces too much normal product, puts product in the wrong place, or produces a damagingly altered product that interferes with the function of the normal gene on the other chromosome. These latter types of dominant mutations are called 'gain-of-function' mutations.

In this book, we are often concerned with the normal, the 'wildtype' function of the gene and its product. Consider the zygotic lethal *fushi tarazu, ftz*. As a zygotic lethal, the phenotype of the mutation depends only on the genotype of the zygote, as distinct from the maternal-effect lethals mentioned above, where the phenotype of the embryo is determined by the mother's genotype. To make it clear whether we are discussing mutant phenotype or wildtype product I use the following terminology: *ftz* or *ftz[+]* means the wildtype gene, as in '*ftz* product' and 'activation of *ftz[+]*'. In this book, an unadorned *ftz* rarely means the mutant, it just refers to the gene. I refer to null mutations as *ftz[-]*, for example in a chromosome deficient for the gene, as in '*ftz[-]* phenotype'.

For more information, see Lindsley and Grell (1968) (details p. 22).

take genes of interest, one by one, and find out when they function. Sometimes the answers are black or white, sometimes grey. To explain, I take four examples.

The first example is the *bicoid* gene; mutant embryos die with a grossly disturbed pattern. Everything depends on the mother's genotype, homozygous *bicoid*⁻ females always lay eggs which produce dead embryos even if the male provides *bicoid*⁺ sperm (for genetic nomenclature see Box 1.1). Clearly, **all** the gene product needed for development is normally deposited in the egg by the mother; it is a true maternal-effect gene.

The second example, *extra sex combs* (or *esc*); this gene was initially defined by a weak mutation that gives viable and fertile flies when homozygous and is recognised because the second and third thoracic legs (T2 and T3; see Box 1.2) are transformed in both sexes towards the pattern of the T1 leg; this is conspicuous in the male because extra sex combs develop, a fairly slight phenotype. However, mothers homozygous for a strong allcle (*esc*⁻), when mated to *esc*⁻ males, produce only embryos that die; these embryos show a complete transformation of all the body segments. In the wildtype, each segment is different, but in these lethal embryos all segments develop alike. Look at Figure 1.1 for the interpretation; at the top of the figure, the mother has *esc*⁺ gene product (shown in red) all over but, importantly, in her germ cells, in the eggs. The male is normal, he has sex comb bristles on the T1 leg only. In the next, the F1 generation, the gene product is present in the egg in sufficient amounts to sponsor development of a normal larva, even though that larva lacks the *esc* gene and can make no gene product of its own. These larvae generate (almost) normal flies — the males have sex comb teeth on all three pairs of legs. However the F1 females have no gene product and they produce eggs with none. If these eggs are fertilised by *esc*⁻ sperm they form mutant embryos. But, if one is fertilised by an *esc*⁺ sperm the embryo develops normally. Clearly the *esc* gene product is a constituent of the normal oocyte, and enough can be provided by either the mother or the father.

> ### Box 1.2 Naming the parts
>
> Because eggs vary in size and pattern themselves in proportion, the position on an egg (in the anteroposterior axis) is conventionally given as per cent Egg Length where 100% Egg Length is the anterior tip of the egg and 0% Egg Length the posterior.
>
> The main segmented part of the body comes from a central region of approximately 68–12% Egg Length, and because the blastoderm is a monolayer of cells, a two-dimensional map of the normal fate of the cells of the blastoderm can be drawn out (see Figure 1.5). These maps do not tell you about the internal states of the cells (whether they are committed to make some structure or not), only how the primordia are disposed.

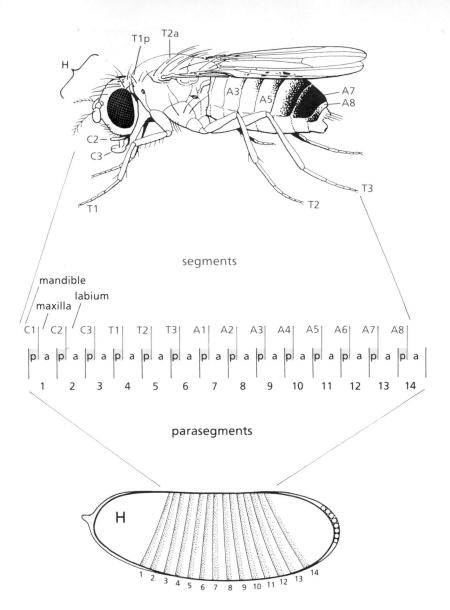

Figure B1.1 The registration of parasegments and segments. There are 14 main parasegments in the trunk of the embryo, plus head (H) and tail regions. The 14 parasegments form the three head segments C1−C3, the three thoracic segments T1−T3 and the eight abdominal segments A1−A8.

The main trunk of the embryo is divided into 14 parasegments (PS 1−14) and this includes ectoderm and the two mesoderms, while the endoderm comes from two regions outside the segmented zone. Foregut and hindgut, which are both named ectoderm, also come from outside the boundaries of the segmented zone.

Parasegments (1−14) and segments (C1−C3, T1−T3, A1−A8) in the larva and adult are offset as shown in Figure B1.1, but this only applies to the epidermis and, possibly, the nervous system. For example, the mesodermal cells of parasegment 4 form most or all the muscle set associated with the wing disc and the second leg (T2), as well as a specific part of the smooth muscle that wraps around the gut. See Chapters 4 and 5.

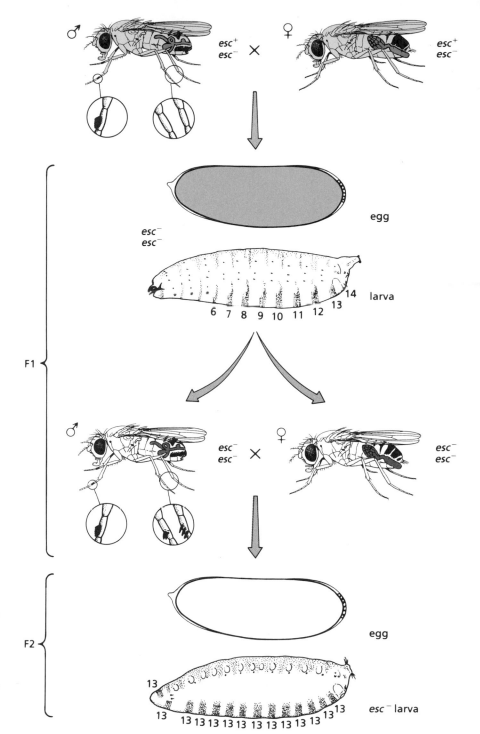

Figure 1.1 The maternal effect of *extra sex combs* (*esc*). Even though the F1 eggs may have no *esc* gene, they inherit sufficient product (indicated in red) from their mother. They pass no product to their offspring (F2) which develop with all their body segments transformed towards the pattern of parasegment 13.

The third example, *Notch⁻*, is a mutation named for its slight dominant phenotype; little notches are taken out from the wing edge in *Notch⁺*/*Notch⁻* flies. Embryos that are homozygous for *Notch⁻* are very badly damaged and never develop to hatching, even if their mothers carry two doses of the *Notch*⁺ gene. So *Notch*⁻ is a zygotic lethal mutation. However, if the mother's germ cells are genetically *Notch*⁻ (see Box 4.1, p. 85) the *Notch*⁻ embryos have an even stronger phenotype. In other words, *Notch*⁻ embryos are partially restored when the mother's germ cells contain the *Notch* gene — it follows that the mother must deposit some *Notch* gene product in the egg, that there is maternal 'rescue' — but not enough to allow the eggs to hatch. The relative contributions of egg and sperm can be assessed in experiments summarised in Figure 1.2. In column II, an egg contains some maternal contribution of gene product, but it is insufficient because if it is fertilised by *Notch*⁻ sperm, development is defective. However a wild-type sperm, plus the maternal contribution, gives enough (column I). Even three doses of *Notch*⁺ in the sperm cannot alone sponsor complete development; in this case the egg derives from *Notch*⁻/*Notch*⁻ germ cells and contains no *Notch* product (column IV). Comparison of the embryos suggests that three doses of *Notch*⁺ in the sperm provides

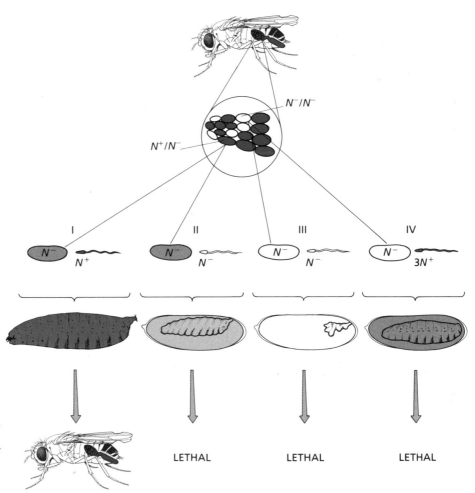

Figure 1.2 Normal development depends on the embryo having enough gene product from the *Notch* (*N*) gene (shown in red). Normally, this comes from both the egg and the sperm. *N*⁻/*N*⁻ oocytes in the mother are made by mitotic recombination (see Box 4.1).

somewhat more gene product than does one dose of *Notch*⁺ in the oocyte.

The fourth example is *engrailed* (*en*), a gene required for segmentation in the embryo. When the phenotypes of *engrailed⁻/engrailed⁻* embryos that have come from *engrailed⁺/engrailed⁻* or *engrailed⁻/engrailed⁻* oocytes are compared, no difference is detected. We can conclude that the product of the *engrailed* gene is not deposited in the egg and the gene is only used in the zygote.

There is another way of looking for maternally derived gene products in the egg: once the gene has been cloned (see Box 1.5, p. 16) a DNA probe for messenger RNA can be prepared and the message looked for in RNA extracted from unfertilised eggs. Although this method is not a test for **function** it is direct and provides a valuable independent check. For example, as might be expected from the genetic experiments, the *bicoid* gene is transcribed in the mother but not in the zygote, *engrailed* is transcribed in the zygote but not in the oocyte, while *esc* and *Notch* are transcribed in both.

We now have information from both genetic and molecular experiments about a sufficient sample of genes and it is clear that about 80% of gene products are present in the egg (see Box 1.3). You might think this suggests that the mother has virtually preformed the embryo in the egg, but not so. The missing 20% includes a crucially important group — gene products that must be present only in a specific subset of cells in the developing embryo. Some of these genes design the embryo, in the sense that the distribution of their products directs the cells to particular fates.

The evidence that there are (at least) two separate groups of genes with respect to expression in the oocyte is quite good. The majority class of genes is needed for building the egg as well as the later stages; these genes will have maternal effects. If they are removed from some germ cells by mitotic recombination or if entirely mutant germ cells are made by transplanting pole cells (p. 35), the embryos are likely to develop poorly or be damaged, or even be non-existent. The minority class of genes is not transcribed during oogenesis and, naturally, their removal from germ cells will be inconsequential. As we have seen, the *engrailed* gene is an example of this minority class. Experiments to be described in Chapter 4 show that this gene is crucial in the allocation of cells to a specific fate. The gene product must not therefore be let loose in the egg — for it could diffuse into the wrong cells and these would be misallocated. It may even be that there is a special mechanism during oogenesis to ensure that genes like *engrailed* are switched off.

The genetic analysis gives a picture of an egg that is too simple; although the mother does not preform the embryo in any detail, she does lay down molecules which pattern the main axes of the body — these are described in Chapter 2.

Box 1.3 Classification and numbers of genes

Most genes are identified by mutations and classified according to the type of mutation. For example a female sterile is a mutation in which females, but not males, are sterile and the gene is often required for some aspect of oogenesis. A zygotic lethal is one in which the wildtype function is required in the zygote. Strictly, it should be sufficient for the wildtype gene to be provided from the father only; to prove this involves making mothers with germ cells that lack the wildtype gene, by mitotic recombination or pole cell transplants (Box 2.2, p. 35). A pupal lethal is a mutation that, when homozygous, results in death during the pupal period.

Maternal rescue always complicates analysis: if the maternal contribution is removed, a pupal lethal mutation may now die in the embryo or a viable female sterile (*esc*, p. 5) may become an embryonic lethal. So when estimates of the number of zygotic lethals are made, say in a defined region of a chromosome, and then these estimates scaled up to give an estimate of the total number of zygotic lethal genes in the genome, it should be remembered that they yield only approximations.

There are other causes of inaccuracy: different mutations in the same gene can have different characteristics. One mutation may be an embryonic lethal, another a viable and this means that estimates of the numbers of genes can go down as well as up. Take the mutation *Ellipse*, a dominant mutation affecting development of the adult eye that was discovered long ago (see Figure 8.5). Dominant but homozygous lethal, it is clearly a gain-of-function phenotype and one might assume its wildtype role to be special to the eye. Another gene *faint little ball* was discovered in a zygotic lethal screen. Still another, *torpedo*, was found in a screen for female steriles; it is a recessive mutation affecting the shape of the follicle and also is a maternal-effect lethal altering the development of the embryo (p. 44). All these mutations have turned out to be alleles of one and the same gene, so what was three genes has become one. The gene encodes a molecule homologous to the receptor for the epidermal growth factor of vertebrates.

The view that genes can be classified according to the type of mutation therefore must be treated with caution, even suspicion. Since it is known that many genes have complex patterns of expression both in time and space and multiple elements of control, both 5' and 3' to the coding region, it is not so surprising that mutations can alter the patterns of expression in a myriad of ways — and produce a fragmented image of the gene and, also, make estimates of the number of genes very tricky.

A mutation that interferes with a process usually tells you the gene is necessary for that process — it does not tell you the gene is sufficient for the process, nor that the gene is exclusively involved in that process — it could be essential for other functions, too. This is particularly relevant with mutations affecting early embryogenesis; the genes may be needed for events later on, but the only requirement for the gene that is conspicuous is the first — once something has gone badly wrong early on, any later roles of the gene become difficult to detect.

The blastoderm and the pole cells

The *Drosophila* egg is, for the first 2 hours, a syncytium — meaning that nuclei divide and migrate in a common cytoplasm. This fact is often pointed out to emphasise how different *Drosophila* is from most other embryos in which the egg cleaves into two and then four complete cells and so on. It is not clear whether this difference is superficial or fundamental; however it is certain that early pattern formation in *Drosophila* takes advantage of the possibility of free diffusion of proteins within the egg, something that is not feasible in many other embryos.

In *Drosophila*, the zygotic nuclei divide very rapidly (every 9 minutes or so) and synchronously for the first seven divisions until, after about 1 hour from egg laying, there are 128 nuclei in the central region of the egg. Most of the nuclei, each with a surrounding islet of cytoplasm, migrate outwards as they continue to divide. Some are left behind; these become the yolk nuclei which divide once, out of schedule with the rest, and are not believed to contribute to the embryo proper. After nine divisions the other nuclei are mostly out near the egg surface and about 15 of these find their way into the posterior pole of the egg. Soon these nuclei become separated into **pole cells**, which begin protein synthesis early and divide on their own schedule about twice more. The individual behaviour of the pole cells and yolk nuclei illustrates the first cellular differentiation in the *Drosophila* embryo.

Why do the pole cells become different from the remainder? The simple answer is that, at the posterior pole, the egg cytoplasm is differentiated before fertilisation and the nuclei that go there respond to this difference. Polar plasm can be transplanted to the anterior pole, and when nuclei enter it there, they form functional germ cells. The polar plasm is sensitive to ultraviolet rays — embryos can be sterilised by irradiating the posterior pole. It includes granules containing RNA, and there is evidence that the RNA is important: the wavelengths of the ultraviolet rays that most easily produce sterility are those absorbed by RNA. Also, if sterilised embryos are irradiated with visible light, fertility can be restored, a process called photoreactivation that occurs only with nucleic acids. It has not been proved that the polar granules themselves are determinative in the sense that their contents instruct the polar nuclei to behave differently — although most believe they are.

As is often the case, understanding comes from studying mutations and the corresponding genes. Mutations in several genes cause a lack of normal polar granules and pole cells; as would be expected these are maternal-effect mutations and their products are constituents of the granules themselves. An example is the *oskar* gene; the *oskar* RNA is present in polar granules and is necessary for the formation of pole cells. Another example is the *vasa* gene. The *vasa* RNA is distributed evenly in the egg, but the protein is translated during oogenesis and much, but not all, finds its way to the posterior pole where it is concentrated. The protein is known, from its sequence, to belong to a family of RNA-binding proteins and is likely therefore to be a key constituent of the polar granules. Females homozygous for mutations in the *vasa* gene produce eggs that lack polar granules and pole cells.

As the pole cells are forming, the remaining nuclei continue to divide in close, but not perfect, synchrony, until they have been through 13 division cycles in all. As they divide, they move out towards the egg surface. What is the pattern of migration? Very different modes of movement can be imagined; at one extreme there could be a precisely oriented first division giving two nuclei, one on the left and one on the right. Each would then divide in a regimented fashion, the left nucleus giving rise to all nuclei that populate the left hand side of the body and the right nucleus to those of the right hand side. The two populations would meet along the midline. At the other extreme, the first cleavage plane could be random and the nuclei could mix up as they divide so that the descendants of the first two, if they were different colours, like salt and pepper, would give a salt and pepper pattern. Fortunately there is a method to label the nuclei, although not always at the first division, and the pattern of migration is known. The method is to make gynandromorphs (flies that are part male, part female) and to map the positioning of the male and female territories (see Box 1.4). In

Box 1.4 Genetic mosaics — gynandromorphs

In *Drosophila* the sex of a cell is determined by its own genotype: one X chromosome for male, two for female. There are several ways of making X chromosomes unstable so that one is lost, occasionally, during mitosis. In a female embryo, loss of an X chromosome from one nucleus or cell will cause that cell to generate a male clone. This chromosome loss may occur at the beginning, at the first mitosis of the zygote, and lead to a mosaic individual with male and female halves. If the lost X chromosome carries a dominant allele that colours the cuticle (such as *yellow*⁺) and the retained chromosome its recessive counterpart (*yellow*⁻) then the result will be a particoloured fly with brown female cuticle and yellow male parts (see Figure 1.3). When loss of the chromosome occurs later than the blastoderm stage, the incipient male cells die because they have problems with dosage compensation (the mechanism whereby gene expression on the **one** X chromosome of males is increased relative to the **two** chromosomes of females so that both sexes have the same amount of gene products). When loss of the X chromosome occurs very late in development, perdurance (see p. 83) of some gene products is sufficient and again small male patches survive.

For more information, see Zalokar *et al.* (1980) and Zalokar and Erk (1976) (details p. 23).

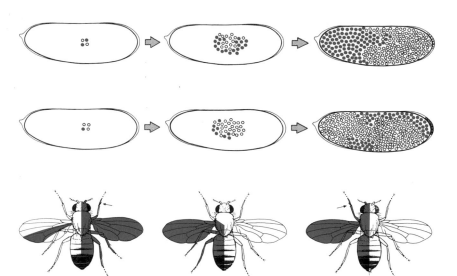

Figure 1.3 Gynandromorphs show that division patterns in the nuclei of the young embryo are irregular. As the nuclei divide (male red, female black) they mix only a little. There is no standard pattern and this causes the larvae and adults to develop with diversely positioned patches of male and female tissue. Sex combs are arrowed.

Figure 1.3 we see that gynandromorphs are mixtures of male (shown in red) and female nuclei. In the top row of eggs, a male nucleus is formed at the first division and in the second row of eggs, at the second division.

The gynandromorphs show that the first plane of cleavage **is** random but the nuclei do not mix up completely as they divide; there is not a salt and pepper pattern, instead there are large male and female territories, but the border between them is wiggly and variable: some cases are illustrated in Figure 1.3. It follows from the variability that there can be no programmatic allocation of cell fate based on a regimented nuclear lineage; instead fate must be assigned to the cells by another mechanism and at a later stage.

Another way of assessing the state of the nuclei is to ask if they are heterogeneous — for example, do the nuclei from different regions have different potentials? If a few nuclei are taken from the central region of a late blastoderm stage and injected into very young embryos of a different genotype, mosaic embryos and flies result. Some of these mosaics are as much as half derived from the transplanted nuclei; occasionally the whole fly is of the donor genotype. These results suggest that all the late blastoderm nuclei are equivalent, or at least unrestricted in which parts they can make (but, remember, they have been removed from their normal environment and transplanted into a new one; a trauma that can produce changes in the nuclei being tested).

Once the nuclei have migrated out to the periphery of the egg, they form a monolayer of cells, the **blastoderm** (stage 5; Figure 1.4). During stage 5, the plasma membrane extends centripetally into the egg and between the nuclei as they become ellipsoidal, eventually cutting in under them to leave, for a time, channels between the nascent cells

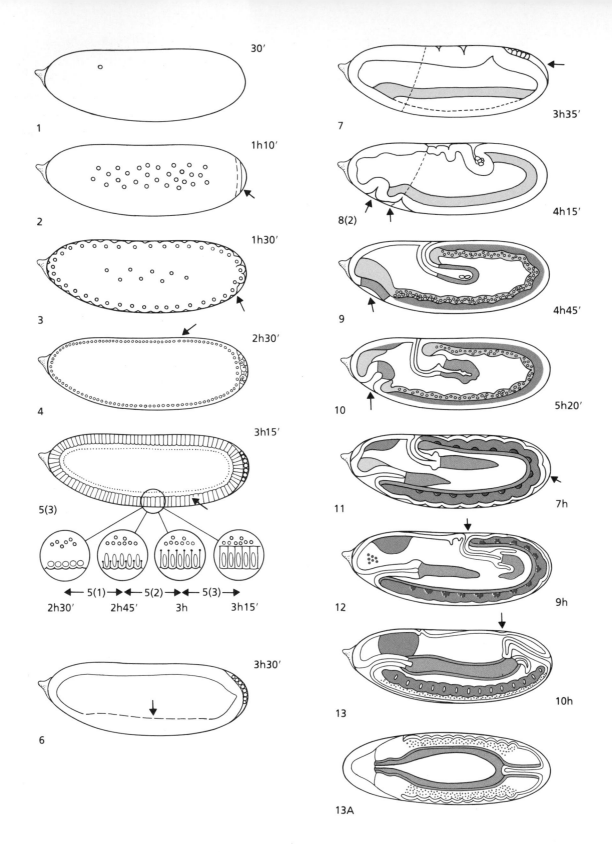

30'

1

1h10'

2

1h30'

3

2h30'

4

3h15'

5(3)

← 5(1) → ← 5(2) → ← 5(3) →

2h30'   2h45'   3h   3h15'

3h30'

6

7   3h35'

8(2)   4h15'

9   4h45'

10   5h20'

11   7h

12   9h

13   10h

13A

and the yolk. Thus, for the whole of the blastoderm period, large molecules such as proteins can pass from one nucleus to the next. So much happens during this long stage (stage 5; Figure 1.4) it is important to subdivide it; this is done by watching the ingrowing membranes. These membranes can be easily seen (see Figure 3.9, p. 70) even in the living embryo and in this way the blastoderm stage, which lasts about 50 minutes, can be divided into three stages of approximately equal duration (5(1)−5(3)).

The early stages of the blastoderm are called syncytial and the late stages are termed 'cellular' blastoderm. In descriptions, no precise distinction is drawn between the two, which is as well, for the cellular blastoderm does not really exist! During the prolonged interphase many things both seen and unseen occur, as the initially homogeneous blastoderm becomes divided up into diverse cell groups, each with a defined role. The result is a ground plan of the embryo, laid out in a two-dimensional cell sheet. As gastrulation occurs (stage 6) the channels pinch off and the cells become separate entities — although it is not known when an unbroken membrane seals the cell off, electrically, from its neighbours. Even after gastrulation, there is evidence from scanning electron microscope pictures that some cells are connected to the yolk and therefore, via the yolk, to each other.

Gastrulation

Gastrulation is a universal and, therefore, important step in animal development. It occurs when a ventral subset of the cells of the blastula roll in together (or dive in one by one) to create a two-layered embryo (stage 6; Figures 1.4 and 1.5). These two epithelia have been traditionally called germ layers, and have become fundamentally distinct. The inner layer is called the mesoderm and the cells in it acquire a distinct potential or 'fate'. Their fate is to form most of the internal organs, such as muscles. The outer germ layer is called the ectoderm, whose cells will make the epidermis and the central and peripheral nervous systems. The concept of a germ layer is built on two separate ideas that are intertwined: the first is concrete; the naming of the cell layer that you see forming itself in the embryo. The second is abstract; the idea that the cells in that layer become allocated to a more specific (and more constrained) fate than before the layer was created. For example,

---

Figure 1.4 (*opposite*) Early development — the stages from 1 to 13. As in nearly all figures in this book, anterior is to left, and ventral down. Ectoderm in white; central nervous system, brown; mesoderm, pink and red; somatic mesoderm, red; visceral mesoderm, pink; and endoderm, grey. 13A shows a horizontal section through stage 13. The arrows indicate features that help staging embryos.

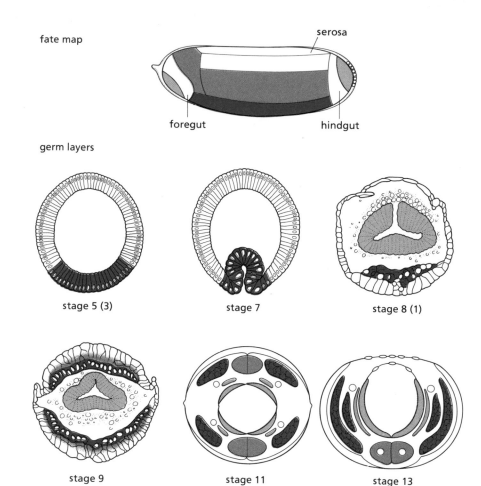

Figure 1.5 Early development — fate map and transverse sections. The fate map shows the disposition of cells on the blastoderm that will, in the normal course of events, generate the parts indicated. Colours as for Figure 1.4. In stage 9 and 11 (the extended germ band) the section cuts across the body twice because the embryo curls back on itself — see Figure 1.4.

fate map

serosa

foregut

hindgut

germ layers

stage 5 (3)

stage 7

stage 8 (1)

stage 9

stage 11

stage 13

in vertebrates, mesoderm cells are considered to be capable of forming muscle, but constitutionally unable to become nerve cells. These two different ideas have often been confounded, and cause and effect have become confused.

I think that genetic criteria should be used to define germ layers and this is beginning to become possible. In *Drosophila* the presumptive mesodermal cells express the *twist* and *snail* genes (pp. 65–6), lose the product of the *engrailed* gene (p. 88) and express the genes of the bithorax complex (Chapter 5) in a pattern that is different from that in the ectoderm. These genes have a controlling role and they and others might determine the special behaviour of the cells in each germ layer.

Molecular and genetic criteria may well lead to a revision of the number of germ layers. An example: the visceral mesoderm has been traditionally included under the general heading of mesoderm. However, it is distinct from the other, the somatic mesoderm. It makes different kinds of cells, the pattern of expression of bithorax complex system of genes within it is unique (p. 126) and it may well have its own control genes. Perhaps, it should be a germ layer in its own right. Similar arguments can be made about the central nervous system.

Mapping of gene expression should help locate the primordia of the germ layers. For example, the cells of the mesoderm invaginate because they have already differentiated from the surrounding cells, as indicated by expression of the *twist* gene. Also, mapping gene expression will help to delineate the boundaries between germ layers, which should achieve resolution at the level of single cells. In Figures 1.4 and 1.5 these different 'germ layers' are shown in various shades: white for epidermis, brown for central nervous system, red for somatic mesoderm, pink for visceral mesoderm, and grey for endoderm.

As the mesoderm cells roll into the furrow, the mass of cells along the ventral midline elongates and pushes around the posterior tip of the egg; as this occurs cells shuffle a little, like a massed crowd entering a narrower walkway, and some neighbours are exchanged. As cells move in ventrally and migrate along posteriorly, most cells move down from the dorsal surface leaving behind a sparse group that never divide again and do not make any part of the embryo itself (Figure 1.5, the serosa).

The extending mass of cells along the ventral midline of the embryo forms the 'germ band', a stage that all insects go through, even though they may get there in different ways. The germ band consists of the main trunk of the future embryo, the part that will become segmented. Inside there is the mesoderm and outside the ectoderm. Before the completion of germ band extension, cells that will form the central nervous system (neuroblasts) dive in individually and these form an intervening layer which can be easily seen in longitudinal and cross sections (brown, Figures 1.4 and 1.5). Meanwhile, two rings of cells, one at the front and one at the back, are beginning to grow inside to form pockets. These are the two rudiments of the midgut, which constitute the endoderm (grey, Figures 1.4 and 1.5); they carry behind them tubes of ectoderm which form the foregut and the hindgut.

After the 13th nuclear cleavage, that is during the whole of stages 5–7, there are no cell divisions, but once gastrulation is completed all the cells in the first of 25 different domains divide. The first five groups divide in fixed positions and in a fixed order in the head and tail, a process which continues until most of the cells of the embryo have divided once. The individual behaviour of these domains, which are arranged in a bilateral and consistent pattern, implies that the cells in one domain are different from those in another. However, even though the division pattern is consistently found in all individuals, it may not be that important. It is possible to tamper with the division patterns: a gene called *string* has been cloned (see Box 1.5), linked to a

Box 1.5 How *Drosophila* genes are cloned

This book is not — could not be — a cloning manual, so I only summarise the principles of cloning. When Morgan chose the fruitfly to work on in 1909, he could not have been aware just how judicious a choice it was. The problem with cloning genes is not usually one of making copies of bits of the DNA and growing those copies up in pure form, but rather in identifying which bit is the gene you want, and being sure about it.

It is instructive to consider how the bithorax complex (BX-C, p. 111) was cloned. First there was the cytogenetics. *Drosophila*'s polytene chromosomes allow the mapping of genes. A nucleic acid probe, labelled with biotin, is applied to spreads of the giant salivary gland chromosomes and the probe attaches to only the corresponding matching sequence, wherever it is in the genome. This marvellous method allows the precise localisation of the gene of interest to the nearest band (Figure B1.2). Although bands vary in size, this can usually place the gene to within 50—100 kilobases of DNA. *Drosophila* has about 200 megabases of DNA, each of the five main chromosome arms being about 40 megabases.

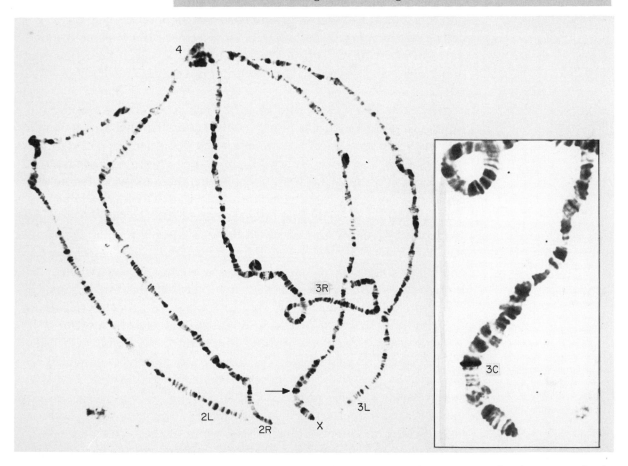

Figure B1.2 The localisation of the gene near *white* to a particular chromosome band. Biotin-labelled probe binds to the complementary DNA sequence at the only place in the genome where it can be found — on region 3C of the first chromosome. Rarely, if ever, have the five chromosome arms arranged themselves so well as in this squash; evidence that at least one scientist is in league with the devil.

Starting with a cloned piece of a transfer RNA that hybridised to bands called 87E1-2, experimenters made use of a fly stock carrying an inversion (a piece of chromosome that is fortuitously cut out by X-rays and reinserts the wrong way round) that took DNA from 87E1-2 and dropped it into the bithorax complex at 89E1-2. They 'walked' — that is stepped along the strand of DNA by finding overlapping sequences — off the end of the inversion straight into the BX-C. Then they went back to wildtype DNA and walked in both directions, collecting overlapping pieces grown up in a lambda phage library until they had a large chunk of continuous DNA extending both sides of their point of entry. They showed they had cloned the right region by analysing (by restriction maps) a large number of BX-C mutations collected by geneticists and finding that some of them made mutually consistent alterations in the DNA map.

Another instructive example is the cloning of the segment polarity gene *wingless* (p. 101). This was first cloned by 'transposon tagging'. Transposons are naturally occurring elements of DNA which are equipped to move about the genome of *Drosophila*, inserting into the chromosomes. They can be mobilised by design and the most useful of these is the P element. If they insert into a vulnerable area of a gene they cause a mutation in that gene. A viable allele of *wingless*, wg^1, was crossed to chromosomes in which P elements had inserted and a new mutation, wg^{CP1}, found that gave a *wingless* mutant phenotype in flies that were wg^{CP1}/wg^1. Labelled probes for the P element were hybridised directly to polytene chromosomes of wg^{CP1}/wg^+ flies and a new insert near to the known site of the *wingless* gene was found. It was likely that the insert was responsible for the mutation in the *wingless* gene and this was nicely confirmed: occasionally wg^{CP1} reverted giving rise to *wingless*$^+$ flies. Every time this happened the reverted flies were found to have lost the P element insert near to the site of the *wingless* gene. The next step was to find the P element in a DNA library made from wg^{CP1} flies; phage-containing P elements were isolated and appropriate DNA from them was hybridised back to the polytene chromosomes. Some were found that were from the *wingless* site and more adjacent sequences were collected by walking.

Because P elements can cause a mutant phenotype if inserted quite far from the gene, there is a real problem in determining which function (identified by finding RNA transcripts homologous to the cloned DNA) represents the gene. One can get help from the genetics — for example, if the gene is a maternal-effect gene one can expect that transcription will occur in the ovary, but proof is still necessary. Two methods are used: the best is to take the piece of DNA including the transcript of interest from the wildtype, and insert it into a P element vector, and transform flies (p. 219). If you have most or all of the gene, the transformed element will rescue, partially or completely, mutations in the gene. If the gene and its controlling elements are very large this experiment may not be possible. Another approach is to make antisense RNA from the transcript sequence and inject that into eggs. In theory, and sometimes in practise, the RNA will bind to the wildtype RNA in the egg, interfere with its function and a mutant phenotype will result. Bits of DNA representing the *Krüppel* and the *wingless* genes were identified in this way.

wingless was also cloned inadvertently. It is often important to look in the fly library for homologues to genes cloned from other animals. This may allow genetic analysis that would be impossible or difficult in humans or other mammals. The *int-1* oncogene from vertebrates had been cloned for some years when a search was made in a fly library for homologous sequences. The gene pulled out by this method turned out to be *wingless*.

Homologies are very useful in cloning and need not represent the entire gene — for example the homeobox sequence (pp. 110 and 216) has been preeminently useful in collecting *Drosophila* genes of interest. In the case of *engrailed* (p. 207) the gene was cloned both by walking and by homeobox homology.

For more information, see Sambrook, J., Fritsch, E.F. and Maniatis, T. (1989) *Molecular Cloning: A Laboratory Manual*. 2nd edn. Cold Spring Harbor Laboratory Press, New York. See also Ashburner (1989) (details p. 22).

heat shock promoter (see Box 2.3, p. 56) and transformed into flies (see Transforming flies, p. 219) — the result is an embryo in which cell division can be induced simply by giving a heat shock. If such an embryo is heat shocked once during the interphase after blastoderm, the *string* gene is ubiquitously expressed and induces a synchronous cell division (the 14th cleavage) which replaces the intricately patterned one. In spite of this, the outcome is a normal fly.

The last germ layer to make its appearance is the visceral mesoderm (pink, Figure 1.4). It arises as a layer on the inside of the mesoderm and wraps around the gut; it is not clear exactly which blastoderm cells generate it, nor whether it shares progenitors with the somatic mesoderm.

Segmentation

Although mapping of gene expression shows that the ground plan of the parasegments (see p. 91) is laid down at about gastrulation, one cannot see segmentation until the germ band has extended. By that stage, the epidermis displays evenly spaced grooves, which demarcate the parasegments (stage 10; Figure 1.4, see Figure 4.4, p. 90) and, inside, the mesoderm is arranged in a series of bulges. The grooves in the epidermis define 14 parasegments and these partially correspond to the three mouth part (G1—G3), three thoracic (T1—T3) and eight abdominal segments (A1—A8). The terminology is devilishly confusing and the reason is simply that parts of the adult were named by segments long before it was realised that the epidermis of the embryo is built on a parasegmental register. The whole business is important, but the reader need not get into a muddle, parasegments and segments do not often need to be correlated and when they do there is always Figure 1.6 and Figure B1.1 to refer to — generally I will use parasegments in the embryo and segments in the adult.

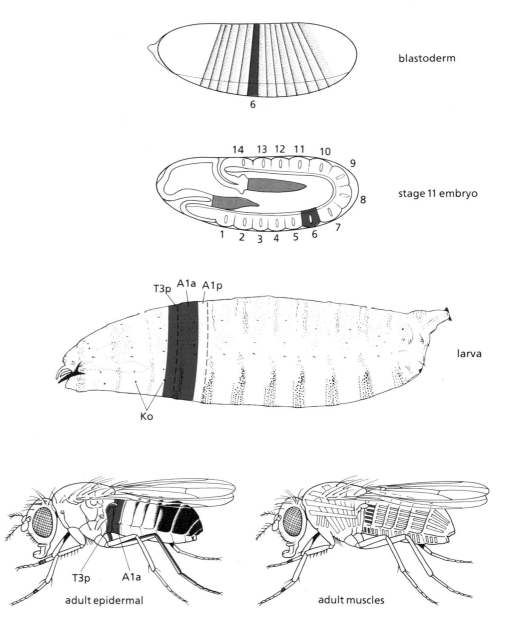

Figure 1.6 The origin and fate of parasegment 6 (red).

Figure 1.6 traces the origin and development of parasegment 6. In the late blastoderm, parasegment 6 is defined as the cells lying between the anterior bounds of two particular stripes of protein made by the *fushi tarazu* and *even-skipped* genes (Chapter 4). By the extended germ band stage, parasegment 6 includes both ectoderm and mesoderm and is demarcated by grooves (below). The cells of parasegment 6 will generate, in the larval epidermis, the posterior compartment of the third thoracic segment (T3p), and the anterior compartment of the first abdominal segment (A1a). The mesoderm cells of parasegment 6 form the muscles that span A1. Still looking at Figure 1.6, the larva is shown with parasegment 6 in red and the nearby segment boundaries by two dashed lines. Note that the anterior boundary of parasegment 6 runs through a thoracic sensillum known as Keilin's organ (Ko). In most of the abdomen the segment borders run just behind the anterior rows of denticles, as shown for the A1/A2 border.

The cells which will make the epidermis, and from which the neuroblasts have segregated, divide two or three times between the blastoderm stage and when they secrete the cuticle at about 14 hours of development. Most of these cells will divide no more; instead they become increasingly polyploid during larval growth and form exclusively larval structures. Small groups of cells remain diploid; these are destined to construct the adult. These cells are selected from all the embryonic cells but are not identified in the blastoderm, meaning they are not chosen until after embryonic cell divisions have occurred and that the descendant of a single blastoderm cell can contribute to parts of both larval and adult epidermis. In the abdomen there are segmentally repeated groups of histoblast cells (approximately 10 in each group) which will replace the moribund larval epidermis in the pupa but do not divide until then. Inside the thorax and genital regions (Figure 1.7) small groups of cells can be identified in the newly hatched larva as the imaginal discs (approximately 40 cells in each one). They grow throughout the larval life and form folded sacs of epithelia that are continuous with the polyploid larval cells. Some of the epithelium is

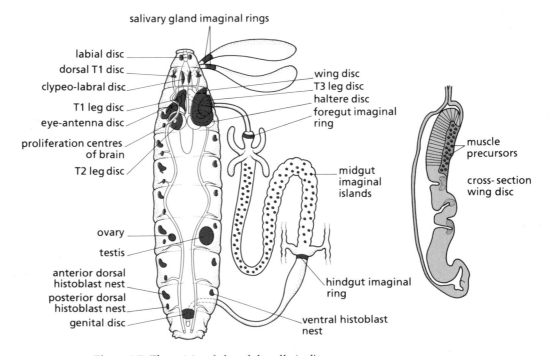

Figure 1.7 The origin of the adult cells (red).

stretched thin, has been called the peripodial membrane and has been traditionally considered to be different from the remaining columnar epithelium. However, it is easier, and probably more correct, not to make this distinction and to think of the disc as a simple sac-like extension of the epithelial sheet, all cells in the epithelium having the potential to make adult cuticle.

The somatic mesoderm forms an uneven epithelium on the underside of the epidermis and, ventrally, it lies in contact with the neuroblasts. How the muscles, fat body and heart arise is poorly understood, but they are segmented structures, implying that the cells of the mesoderm are not equivalent but are also divided into segmental sets, each set being different from the next. Larval muscles arise in a very precise

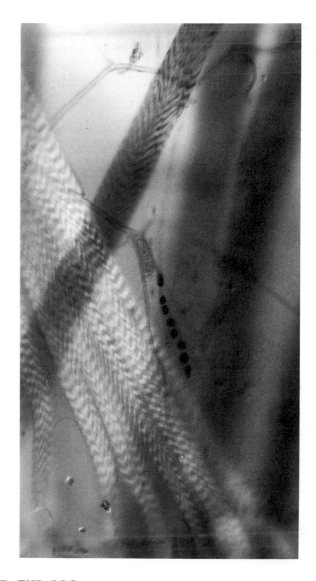

Figure 1.8 The source of the adult muscles. In amongst the larval muscles and associated with specific nerves are rows of small cells that will divide to construct the adult muscles — they express the *twist* gene.

pattern, individual muscles being first detectable as single large founder cells to which other mesoderm cells fuse. As in the case of the ectoderm, it seems certain that specific cells are kept back to form the adult; although these are hard to find in the young larva, they turn up later on inside the imaginal discs as 'adepithelial cells' and these make the adult muscles. As antibodies to gene products become available they can be used to pick out classes of cells that otherwise melt into the crowd. For the precursors of the adult abdominal muscles, the anti-*twist* antibody has proved very useful. The *twist* gene is involved in the specification of embryonic mesoderm and is later expressed in small sets of cells that are associated with particular larval muscles. These cells proliferate during the larval period and form little chains or groups of cells associated with nerve bundles or trachea (Figure 1.8). Later each group forms a specific muscle in the adult.

Pattern formation in the embryo, both in the ectoderm and meso-derm, results in the accurate deployment of groups of cells, each being placed correctly and each being given a specific morphogenetic task. The next chapters describe experiments that begin to explain how this is achieved.

Further reading

REVIEWS

Ashburner, M. (1989) *Drosophila: A Laboratory Handbook*. Cold Spring Harbor Laboratory Press, New York.
Ashburner, M. and Novitski, E. (eds) (1976) *The Genetics and Biology of Drosophila*. Vol. 1. Academic Press, New York.
Ashburner, M. and Wright, T.R.F. (eds) (1980) *The Genetics and Biology of Drosophila*. Vol. 2. Academic Press, New York.
Campos-Ortega, J.A. and Hartenstein, V. (1985) *The Embryonic Development of Drosophila melanogaster*. Springer, Berlin.
Demerec, M. (ed.) (1950) *Biology of Drosophila*. John Wiley and Sons, New York.
King, R.C. (1970) *Ovarian Development in Drosophila melanogaster*. Academic Press, New York.
Lindsley, D.L. and Grell, E.H. (1968) *Genetic Variations of Drosophila melanogaster*. Carnegie Institute of Washington Publication No. 627. (This is an invaluable catalogue of genes and mutations. It is currently being updated: see *Drosophila Information Service*, which is an irregular journal, full of lab lore.)
Roberts, D.B. (ed.) (1985) *Drosophila: A Practical Approach*. IRL Press, Oxford.
Sander, K. (1976) Specification of the basic body pattern in insect embryogenesis. In: Treherne, J.E., Berridge, M.J. and Wigglesworth, V.B. (eds) *Advances in Insect Physiology*. Vol. 12, pp. 125–238. Academic Press, New York.

SELECTED PAPERS

Early development

Foe, V.E. (1989) Mitotic domains reveal early commitment of cells in *Drosophila* embryos. *Development* **107**: 1–22.

Zalokar, M. and Erk, I. (1976) Division and migration of nuclei during early embryogenesis of *Drosophila melanogaster*. *J. Microscopie Biol. Cell.* **25**: 97–106.

esc

Frei, E., Baumgartner, S., Edström, J.-E. and Noll, M. (1985) Cloning of the *extra sex combs* gene of *Drosophila* and its identification by P-element-mediated gene transfer. *EMBO J.* **4**: 949–987.

Struhl, G. (1981) A gene product required for correct initiation of segmental determination in *Drosophila*. *Nature* **293**: 36–41.

Gastrulation

Leptin, M. and Grunewald, E. (1990) Cell shape changes during gastrulation in *Drosophila*. *Development* **110**: 73–85.

Germ cells

Wieschaus, E. (1978) The use of mosaics to study oogenesis in *Drosophila melanogaster*. In: Subtelny, S. and Sussex, I.M. (eds) *The Clonal Basis of Development*. Academic Press, New York.

Gynandromorphs

Zalokar, M., Erk, I. and Santamaria, P. (1980) Distribution of ring-X chromosomes in the blastoderm of gynandromorphic *D. melanogaster*. *Cell* **19**: 133–141.

Zusman, S.B. and Wieschaus, E. (1987) A cell marker system and mosaic patterns during early embryonic development in *Drosophila melanogaster*. *Genetics* **115**: 725–736.

Larva

Bryant, P.J. and Levinson, P. (1985) Intrinsic growth control in the imaginal primordia of *Drosophila*, and the autonomous action of the lethal mutation causing overgrowth. *Dev. Biol.* **107**: 355–363.

Lohs-Schardin, M., Sander, K., Cremer, C., Cremer, T. and Zorn, C. (1979) Localized ultraviolet laser microbeam irradiation of early *Drosophila* embryos: fate maps based on location and frequency of adult defects. *Dev. Biol.* **68**: 533–545.

Maternal contribution

Garcia-Bellido, A. and Robbins, L.G. (1983) Viability of female germ-line cells homozygous for zygotic lethals in *Drosophila melanogaster*. *Genetics* **103**: 235–247.

Lawrence, P.A., Johnston, P. and Struhl, G. (1983) Different requirements for homeotic genes in the soma and germ line of *Drosophila*. *Cell* **35**: 27–34.

Perrimon, N., Engstrom, L. and Mahowald, A.P. (1984) The effects of zygotic lethal mutations on female germ line functions in *Drosophila*. *Dev. Biol.* **105**: 404–414.

Muscle development

Bate, M., Rushton, E. and Currie, D.A. (1991) Cells with persistent *twist* expression are the embryonic precursors of adult muscles in *Drosophila*. *Development* **113**: 79–89.

Notch

Jiménez, F. and Campos-Ortega, J.A. (1982). Maternal effects of zygotic mutants affecting early neurogenesis in *Drosophila*. *Wilhelm Roux's Archives* **191**: 191–201.

oskar

Lehmann, R. and Nüsslein-Volhard, C. (1986) Abdominal segmentation, pole cell formation, and embryonic polarity require the localized activity of *oskar*, a maternal gene in *Drosophila*. *Cell* **47**: 141−152.

Parasegments

Martinez-Arias, A. and Lawrence, P.A. (1985) Parasegments and compartments in the *Drosophila* embryo. *Nature* **313**: 639−642.

string

Edgar, B.A. and O'Farrell, P.H. (1990) The three postblastoderm cell cycles of *Drosophila* embryogenesis are regulated in G2 by *string*. *Cell* **62**: 469−480.

vasa

Lasko, P.F. and Ashburner, M. (1988) The product of the *Drosophila* gene *vasa* is very similar to eukaryotic initiation factor-4A. *Nature* **335**: 611−615.

SOURCES OF FIGURES

For details, see above.
Figure 1.1 See Struhl (1981).
Figure 1.2 See Jiménez and Campos-Ortega (1982).
Figure 1.3 See Zalokar *et al.* (1980) and Zusman and Wieschaus (1987).
Figure 1.4 After Roberts (1985), see also Campos-Ortega and Hartenstein (1985).
Figure 1.5 After Demerec (1950) and Leptin and Grunewald (1990).
Figure 1.6 See Martinez-Arias and Lawrence (1985). Larva from Lohs-Schardin *et al.* (1979).
Figure 1.7 After Bryant and Levinson (1985).
Figure 1.8 Photograph courtesy of M. Bate and E. Rushton.
Figure B1.2 Photograph courtesy of M. Ashburner.

2 The first coordinates

THE MOTHER SETS UP four independent systems in the egg: (1) the anteroposterior gradient of bicoid *protein which derives from RNA that is localised at the anterior pole; (2) a posterior system which is necessary for the formation of the abdomen, whose effective agent is the* nanos *gene; (3) a terminal system which is responsible for defining the head and tail of the embryo and depends on localised activation of a receptor protein (*torso*) at the ends of the egg; (4) a dorsoventral system that depends on activation of a receptor protein (*Toll*) along the ventral midline of the egg.*

This chapter is concerned with analysis of the first steps of development, which means that we start with the mother's contribution to the egg. The egg is an asymmetric structure and this is defined during oogenesis. First, the germ cell divides to produce 16 cells that are incompletely separated from each other; cytoplasmic channels link them all together. One of the most posterior cells becomes the oocyte and stays diploid and the others become the polyploid nurse cells; they synthesise material which is pumped into the growing oocyte at what will be the anterior end of the embryo. The little group of 16 cells is encapsulated in a bag of about 1000 small follicle cells that derive not from the germ cells but from the mesoderm of the mother. The oocyte sits in the chamber such that the future posterior end of the embryo leads outwards and the more convex surface of the egg becomes the ventral part of the embryo.

The shape of the egg only correlates with the axes of the embryo, it does not itself determine the axes — for ingenious experimentalists can cause the making of a head where the tail normally appears or dorsal where it should be ventral. Sander showed that the dorsoventral and anteroposterior axes are not fixed at the same times and respond differently to experiments (Figure 2.1). He concluded that the two axes are determined by independent systems and that they work in different ways. The dorsoventral and anteroposterior systems also have different jobs to do. In the anteroposterior axis there is a head and a tail and a series of reiterated elements or segments, while in the dorsoventral axis there is a single succession of cells of different types and no reiteration (p. 14). We shall therefore look at the two systems in turn.

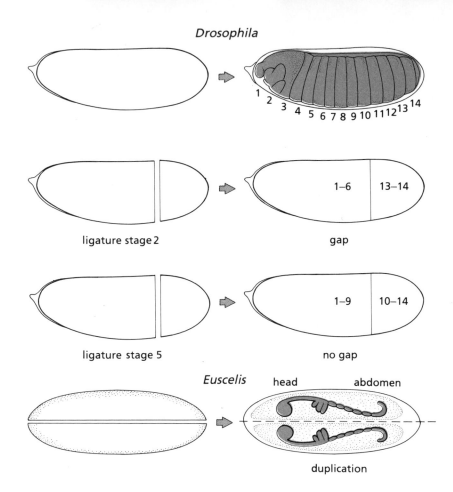

Figure 2.1 Sander's experiments on the determination of the main body axes.

The anteroposterior gradient (see The morphogenetic gradient, p. 204)

Although Sander worked mainly with a leaf-hopper called *Euscelis* and not *Drosophila* his experiments are important, and, in essence, they probably apply to *Drosophila* as well. He found a way of ligating the leaf-hopper egg in half such that both halves could develop; he attached a razor blade to a machine that lowered it very steadily and so slowly that it did not cut, but deformed the egg until it was effectively divided. Each half formed its own separate blastoderm and went on to develop independently. If this was done late in development after gastrulation (see p. 13) each half made the parts it would normally make — showing that the operation did not itself produce a zone unable to develop near the cut. However when the operation was done earlier, the anterior and the posterior halves produced a few enlarged anterior segments and posterior segments respectively — the middle segments were missing. The earlier the egg was cut, the bigger the missing 'gap' of middle segments.

As shown in Figure 2.1, ligating the egg at stage 5 produces two half-embryos whose parts together add up to the whole pattern. Ligating the egg earlier, at stage 2, results in a large gap in the pattern that corresponds approximately to parasegments 7–12. What happens is not the death of the cells that should make the middle of the embryo, but their incorrect specification — the whole ground plan of the embryo alters and cells that would have made one part of the body now make another. Sander suggested a nice explanation: he proposed that there are anterior and posterior factors localised at the poles of the egg, and during early development these normally produce graded influences that spread from the poles and interact in the middle of the egg — an interaction that is prohibited by the ligature. 'The simplest model to account for the results assumes some kind of morphogenetic gradient to which nuclei or cells react differently according to local level.' [5]

He also did some other eloquent experiments. At the posterior pole of the leaf-hopper egg, there is a mysterious blob (black dot in Figure 2.2) which actually contains symbiotic organisms that are being handed on from mother to offspring. This blob provides a marker for posterior pole cytoplasm and Sander was able to move it and its associated posterior cytoplasm around in the egg. If it is transposed up to the middle of the egg and the egg cut behind it, a whole embryo develops in the anterior fragment and extra segments form in the posterior fragment (Figure 2.2 I). The posterior polar cytoplasm does not always lead to a complete embryo in the anterior fragment, sometimes its influence is only sufficient to raise the level of the pattern partially (Figure 2.2 I). This shows that the posterior cytoplasm is not a simple determinant for the most posterior segments but can have a **quantitative** effect, pushing cells near to it in a more and more posterior direction. In Figure 2.2 II the ligature is placed anterior to the transplanted polar material; the polarity of the partial embryo that then forms can be completely, or partially, reversed. It is effects like these that led Sander to his gradient model, which we can now restate as follows: there are, in the newly laid egg, two centres, one at each pole, which produce spreading influences that interact in a graded fashion along the egg axis. The relative amounts of these two influences at each point determine the local pattern — in other words, the scalars of the gradients determine the segment type and the slopes determine the order and polarity of the segments. Note that the model is a general one; it does not specify the detailed mechanism. The gradients might be concentration gradients of molecules, and the concentration landscape with its local scalars and vectors could prefigure the pattern of differentiated cells and their polarity. Molecules whose local concentration determines the local pattern of differentiation in this way have been called **morphogens**. Detailed models of this type have been developed to explain pattern formation within the insect segment; they are discussed in Chapter 6.

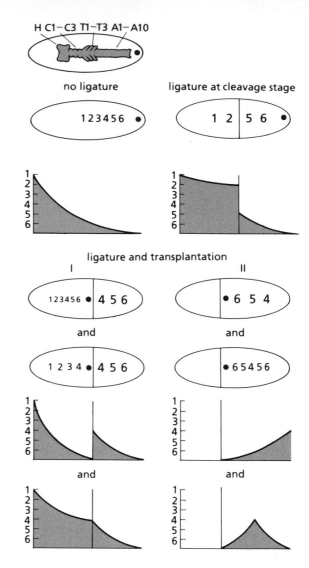

Figure 2.2 Evidence for an anteroposterior gradient — experiments on *Euscelis*. The body pattern of the embryo is indicated by the sequence and size of numbers 1–6, and the gradient interpretations shown on the same scale.

Similar egg-cutting experiments have been done with *Drosophila* and, just as in the leaf-hopper, when the body parts made by the anterior and posterior fragments are added up, there is a gap in the middle segments when the cut is made early, but not when it is made late. Comparable transplantations of cytoplasm have been done in *Drosophila*, but first some genetics.

Genes and the anteroposterior pattern

Amongst the products of the screen for maternal segmentation genes (p. 201) was the *bicoid* gene. Look at Figure 2.3: embryos developing from mothers that lack the gene are normal in the posterior pattern, have disrupted anterior abdominal segments and no proper

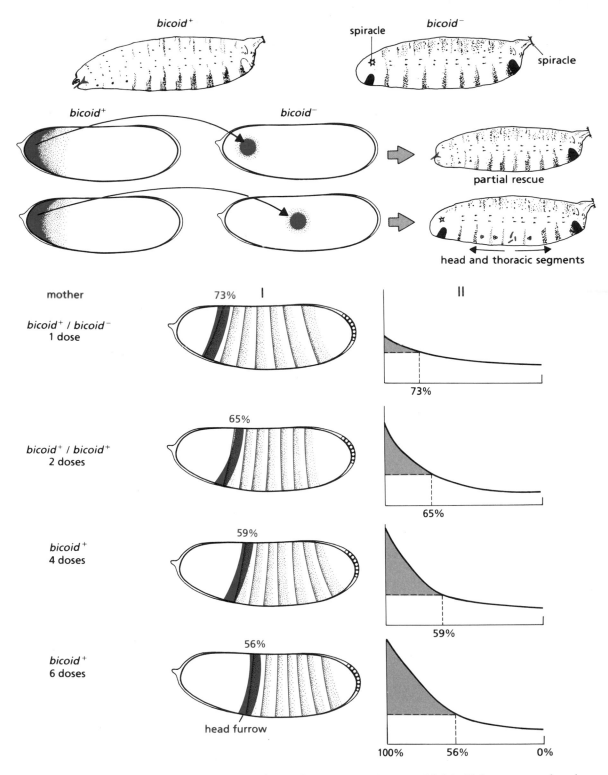

Figure 2.3 Dosage and transplantation experiments with *bicoid* demonstrate that the *bicoid* protein acts as a morphogen.

head and thorax. In place of the head there is an extra telson (shown in black) and spiracles, structures normally found at the posterior end of the embryo. Cytoplasm from *bicoid*⁺ eggs can be injected into *bicoid*⁻ hosts and this can rescue the mutant phenotype to various degrees (Figure 2.3). The more doses of *bicoid*⁺ there are in the mother, the more complete the rescue; moreover the more anterior the source of the cytoplasm, the more effective it is. These observations are reminiscent of Sander's experiments; they show that *bicoid* activity can have quantitative effects.

Further, there is a quantitative relationship between the number of *bicoid*⁺ genes in the mother and the pattern of the embryo. As the dose increases, the embryo gains anterior pattern at the expense of posterior. Figure 2.3 shows this in two ways; first, there is the shifting of an early feature in the pattern, the head fold (red), which moves further and further back with increasing numbers of *bicoid*⁺ genes. It is normally at 65% Egg Length (see Box 1.2), but when the mother has only one dose of *bicoid*⁺, the size of the head is reduced and the head fold moves anteriorly to 73% Egg Length. As doses of *bicoid*⁺ in the mother are increased one by one to 6, the position of the head moves back stepwise to 56% Egg Length. Second, the ground plan of the embryo, in the anteroposterior axis, is illustrated by the layout of the *fushi tarazu* stripes (p. 96). As the doses of *bicoid*⁺ are increased, so these stripes become more and more crowded into the posterior region of the egg. Finally, as shown in Figure 2.3, if anterior cytoplasm from *bicoid*⁺ eggs is placed in the middle of *bicoid*⁻ embryos, then head (red) and thoracic structures can develop **there** — suggesting that the *bicoid* product itself is the patterning agent, and is homologous to the anterior influence described by Sander; in short it is a morphogen. The effect of these transplantations appears to spread from the site of the operation and set up new and abnormal patterns that display unusual polarity.

The *bicoid* gene has been cloned and sequenced (it has a special group of 180 base pairs called a homeobox, p. 110); one of the first questions to ask about it is, where is the maternal RNA? — remember it is a maternal-effect mutation. There are two ways of finding out where RNA is, and both methods are types of hybridisation *in situ*. First, radiolabelled probes can be hybridised to sections; they bind to homologous RNA sequences wherever they are. The sections can be dipped in emulsion and the sources of radioactivity localised. In the second method, the probe is labelled with a chemical (digoxigenin) that can be revealed by a coloured stain. Using these methods the *bicoid* RNA was traced to the nurse cells (where it becomes localised within them) and to the oocyte, where it is found exactly where it should be — at the anterior pole of the egg. The *bicoid* RNA is shown in red in Figure 2.4.

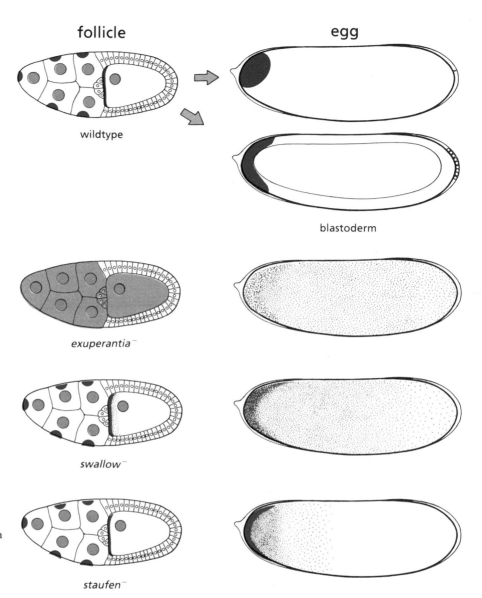

follicle   egg

wildtype

blastoderm

exuperantia‾

swallow‾

staufen‾

Figure 2.4 The localisation of *bicoid* RNA in the follicle and of the egg of wildtype and various mutants.

How does the RNA get to the correct destination and what holds it in the right place? It is thought that a particular sequence in the messenger RNA is recognised by a binding protein and this protein is itself anchored in the egg cytoplasm near the membrane. If a truncated *bicoid* gene is transformed into flies (Transforming flies, p. 219), it can produce a message that fails to localise; using various deletions it is therefore possible to find which part of the messenger RNA is necessary for localisation. The answer for the *bicoid* gene is all or part of the 3' untranslated region — most likely this forms a three-dimensional structure which presents a particular shape or sequence that is recognised by the binding protein. So far three genes that are necessary for localisation of the RNA have been identified; *swallow, exuperantia* and *staufen* (see Figure 2.4 for their effects on the *bicoid* RNA). It is not yet clear what these genes do. Embryos from *staufen⁻* mothers are defective at both ends; indeed, the *staufen* protein is concentrated at both ends of the egg and may associate with *bicoid* and *oskar* RNA.

The identification of *bicoid* is important because it is the first morphogen to be discovered and it plays a crucial part in the patterning of the egg. Antibodies (see Box 2.1) against the *bicoid* protein show it is absent from eggs when they are laid but soon appears, close to the RNA which encodes it. By the time pole cells are being formed it is present in a concentration gradient that extends from 100% to about 30% Egg Length, far beyond the RNA source from which it presumably spreads by diffusion. The protein has a short half life, probably less than half an hour — otherwise the egg would just fill up with protein and the gradient would be obliterated.

The gradient landscape of concentration prefigures and determines the pattern of the embryo. This is shown by varying the amount and

---

Box 2.1 Antibodies

Antibodies are being used more and more in developmental genetics. A specific antibody against the whole or part of a protein is valuable because it can be used to locate a gene product in the cell or in the embryo. The antibody is made by mass producing all or part of the protein in bacteria, usually using a hybrid gene in which DNA from the gene of interest is spliced into the *lacZ* gene of *E. coli* (which encodes the protein β-galactosidase). This results in the production of a hybrid protein which is purified and injected into a mammal such as a rabbit. The rabbit produces diverse antibodies some of which are against the gene product of interest, and these can be purified.

There are various ways of locating antibodies in embryos but one of the most common is as follows: you have an antibody against a protein of interest, such as the *bicoid* gene product, that was raised in (say) a rabbit. You wash fixed embryos in a low concentration of the antibody and the antibody binds to the *bicoid* protein in the embryo. After washing away the unbound excess you then treat with a secondary antibody that recognises all rabbit antibodies, and this antibody is chemically linked to biotin, or is fluorescent. The secondary antibody binds to the rabbit anti-*bicoid* antibody and can be seen either by locating biotin with a coloured stain, or by looking for fluorescence.

For embryos there is the problem of getting fixatives, antibodies and other reagents into the embryo, past the highly impermeable vitelline membrane. This problem has been overcome; octane and heptane punch small holes in the vitelline membrane and let in the reagents.

Antibody stainings for proteins give higher resolution than the *in situ* methods for RNAs and it is also very useful since two antigens can be mapped in one embryo with different dyes, allowing the expression pattern of one gene to be compared directly with another.

distribution of the *bicoid* RNA in the mother and consequently the scalar value of the protein gradient. Once again genetics is used to do this. Look again at Figure 2.3 II. The maximum concentrations and shapes of decline of different *bicoid* protein gradients are estimated by direct measurements on the eggs and these are compared with the observed positions of the head folds. They show a reasonably good fit — although not as good as Figure 2.3 II suggests, where the fit has been exaggerated so that the principle is understood.

The discovery of the *bicoid* gradient leads us on, for the next question is 'how is the *bicoid* concentration interpreted to give pattern?' This question is approached on p. 51 — before that we have to finish building our picture of the unfertilised egg.

Posterior pattern

There are about nine maternal-effect mutations in which the posterior part of the body pattern is defective; they lack the abdominal segments altogether and look extremely dwarfish (see Figure 2.6). Two examples are *oskar* (*osk*) (the dwarf in *The Tin Drum*) and *nanos* (*nos*) (a mutant version of the Spanish for dwarf). It was at first thought that one of these genes would be like *bicoid*, that is it would produce a morphogen gradient with a high posterior limit that declines anteriorly, as suggested by Sander's experiments. But, as we shall see, this did not turn out to be quite right.

When there is a family of genes with similar mutant phenotypes it is important to determine if there is one gene that is primary, meaning that it does something to the embryo, that it acts on a process or has an effect on subordinate genes that have a function at a later stage. The other genes in the family may be auxiliary, necessary for the function of the primary gene but not specific for it; they may be needed for other, different processes as well. The subordinate genes which function later are called 'downstream' to emphasise that their action is controlled by, or dependent on, the earlier action of the active 'upstream' gene.

So, which is the primary gene in the posterior group? As with *bicoid*, rescue experiments were done, but now cytoplasm was taken from the nurse cells and injected into *nanos*⁻ mutant eggs. Rescue of the mutant phenotype shows that the cytoplasm of the nurse cell contains the 'posterior activity'. This posterior activity was found to be present in the nurse cells of all but one of the mutants — implying that these mutations do not remove the activity itself but interfere with its translocation to the posterior pole of the egg or with its packaging or function. However, *nanos*⁻ nurse cells lack all posterior activity and it was concluded that the *nanos* gene product is the posterior factor. Even synthetic *nanos* RNA will rescue eggs mutant for all known posterior group genes. *nanos* is therefore the primary gene; but also the logic of these experiments means that it is formally the last gene in the pathway, the most 'downstream'.

The *nanos* gene has been cloned and indeed the RNA is tightly localised to the extreme posterior pole of the egg (Figure 2.5). One imagines it is translated like the *bicoid* RNA and a concentration gradient of protein spreads out from the localised RNA. The wildtype

Figure 2.5 The *nanos* RNA is tightly localised to the posterior pole of the egg.

function of *nanos* is not equivalent to that of *bicoid* however: the embryo does not appear to respond point by point to the local concentration of *nanos* protein. Although transplantation of *nanos* RNA into the anterior end of the egg can induce posterior structures there, it works by interfering with the translation of *bicoid* RNA. The impression given is that *nanos* is **permissive**, it allows posterior pattern, but is not **determinative**, as *bicoid* is.

To understand how *nanos* works we must meet another gene, called *hunchback*. The *hunchback* gene has primarily a zygotic role and mutants have a lethal phenotype much like *bicoid⁻*, lacking anterior parts. There is also some *hunchback* RNA provided by the mother, and this is at first distributed evenly. Strangely, this *hunchback* maternal RNA is dispensable — *hunchback⁻* germ cells which contain no gene product (see Box 2.2) can form normal embryos when the eggs are fertilised by *hunchback⁺* sperm. For normal development, *hunchback* protein must be in a gradient, high anterior and low posterior; this is achieved in two ways: first, the ubiquitous maternal RNA is translated only at the anterior end of the egg, and second, the zygotic RNA is made only at the anterior end.

If *hunchback* RNA is translated in sufficient amounts in the posterior region of the egg — for example, as happens when the *hunchback* gene is driven by a heat shock promoter — the *hunchback* protein disrupts abdominal development and the deformed embryos resemble *nanos⁻* ones. The explanation is as simple as it is unexpected: the role of the *nanos* gene must be to disable *hunchback* maternal RNA at the posterior end of the egg. The proof follows logically: if the only role of the *nanos* gene is to stop the *hunchback* maternal RNA from being translated in the posterior half of the egg, embryos without maternal *hunchback* RNA should not need the *nanos* gene either. Consequently, eggs lacking the maternal contributions of both the *hunchback* and *nanos* genes should develop normally when fertilised by *hunchback⁺* sperm. They do. The experiments are explained in Figure 2.6. In A, the wildtype situation, the *nanos* gene product clears the *hunchback* RNA from the posterior region allowing abdomen development there. In B, without

Box 2.2 Genetic mosaics — pole cell transplantation

This is easy to understand but tricky to do. The purpose is to study embryos that lack all product of the gene of interest and for many genes this can be difficult, since the mother deposits gene product into the embryo and even homozygous mutant zygotes have some product. Also, the mutation will often be lethal, so homozygous mutant mothers cannot be made. The solution is to have a normal fly with germ cells that are homozygous for a mutant in the gene of interest and this is done by transplanting pole cells between embryos. Female embryos that carry a particular dominant female sterile mutation (*Dfs*, see Fig. B2.1) and therefore can make no competent

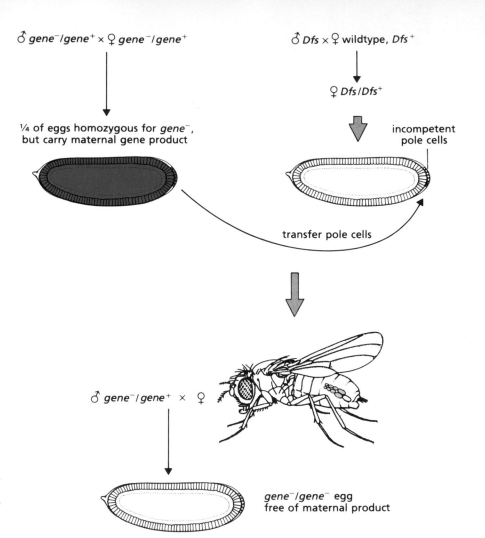

$\male\ gene^-/gene^+ \times \female\ gene^-/gene^+$

$\male\ Dfs \times \female$ wildtype, Dfs^+

$\female\ Dfs/Dfs^+$

incompetent pole cells

¼ of eggs homozygous for $gene^-$, but carry maternal gene product

transfer pole cells

$\male\ gene^-/gene^+ \times \female$

$gene^-/gene^-$ egg free of maternal product

Figure B2.1 Pole cell transplantation can be used to make germ cells that lack a particular gene; eggs deriving from those cells will have no gene product.

pole cells of their own, are used as surrogate mothers and embryos homozygous for the lethal in question are used as donors. Pole cells are taken up intact into a micropipette from one donor and put into the posterior pole of the hosts. These pole cells then grow up into germ cells in the ovary of the surrogate. Her own pole cells fail to develop and consequently all the eggs she lays derive from the transplanted pole cells (Figure B2.1). Success will normally be limited to one in 16 of transfers, for there must be a female donor and a female host, and, usually, only one in four of the donors will be homozygous for the mutant of interest.

These experiments can be used to assess the maternal contribution of a wildtype gene to the egg (p. 6) and to ask whether mutant phenotype depends on germ cells or soma — in these surrogate mothers only the germ cells are mutant, the follicle cells derive from her soma, which is not altered by the experiment.

For more information, see Roberts (1985) (details p. 22).

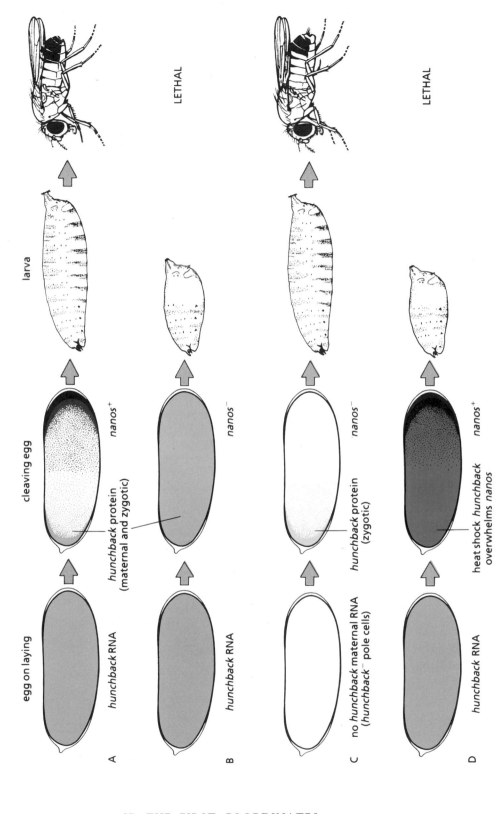

Figure 2.6 The main function of *nanos* is to block the *hunchback* RNA in the posterior region of the egg. If no maternal *hunchback* RNA is made, *nanos* is dispensable.

the *nanos* gene product, *hunchback* RNA is translated into protein and prohibits abdomen development. In C, the *hunchback* gene is removed from the germ cells, there is no *hunchback* maternal RNA, *nanos* becomes redundant, and the embryos form perfect flies. In D, the *nanos* function is swamped by expressing *hunchback* under a heat shock promoter and *hunchback* protein is made at the posterior end, inhibiting the development of abdomen.

There is a little army of genes that are necessary to ensure *nanos* RNA works; that the oocyte and nurse cells are made correctly (e.g. *egalitarian*), that the RNA is localised to the posterior pole (*Bicaudal-D*), packaged properly when it gets there (*oskar, staufen, valois, vasa*) and is deployed correctly (*pumilio*). As would be expected, mutations in these genes are maternal-effect, it being the mother's role to construct the egg in all its parts. Of course, the observation that these genes are upstream of *nanos*, does not mean that their only or main role is concerned with *nanos* localisation or proper function, it only means that the *nanos* gene cannot do its job unless they have done theirs.

It seems perverse that the fly should have a specific genetic system to clear up the unwanted expression of another gene. Would it not be simpler to switch off *hunchback* in the mother? This result illustrates a general principle and delivers a chastening lesson; the principle is that evolution works on what is there, it tinkers, it does not look at the whole system and devise a logical or economical solution. The lesson tells us that trying to understand embryos by using logic and theory is risky or impossible. It is better to do experiments and find out through those. The next part of our story is a good example of this, because none of the theoreticians predicted a special group of maternal-effect genes concerned with making the extreme anterior and posterior ends of the embryo.

The terminal system

These new genes are maternal-effect; homozygous mutant mothers lay eggs in which embryos develop with missing anterior and posterior extremities, that is the acron at the head end and the telson at the tail. There is a group of genes each having a similar mutant phenotype. The *torso* gene has been well studied; it has been cloned, its sequence suggests a tyrosine kinase receptor and it appears on the plasma membrane of the egg during the cleavage stage, where it is distributed evenly. One would guess that a receptor would bind a ligand, and the ligand might be present only where activated receptor might be wanted — at the ends of the eggs. There are dominant mutants of *torso* in which the embryos have enlarged terminal regions and little in between; it seems very likely that, in these, the receptor protein is altered in such a way that, even in the absence of the ligand, it is constitutionally active. The various genes of the terminal system are collected and ordered in a **pathway** of action by epistasis studies (see Box 2.3). Genes encoding for, or modifying, a ligand should have mutant phenotypes similar to *torso*$^-$ mutants but be upstream of the *torso* gene. When the tyrosine kinase receptor binds a ligand, this must be transduced into the cell; there will be genes involved here and in receipt of the signal inside the cell, these will be downstream of the *torso* gene. Consider an example, *trunk*, a maternal-effect gene which we want to place in the pathway. The argument uses double mutants, females which are *trunk*$^-$ and also carry a dominant allele of *torso*. If *trunk* acts upstream of *torso*, the double mutant will still show the dominant *torso* phenotype because this dominant allele is expected to be independent of functions normally before it; while if it is downstream it will show the *trunk*$^-$, that is the *torso*$^-$ phenotype because the dominant allele could not produce its final effect on pattern without the later functions in the pathway. The *trunk* gene proved to be upstream, as did *torsolike*, another maternal-effect gene.

Now, most genes that determine pattern are germ cell dependent; which means that when the ovary of the mother is a genetic mosaic, the germ cells being of one genotype (say, mutant) and the mesoderm follicle of another (wildtype), the eggs make embryos that correlate with the genotype of the germ cells not of the follicle cells (see p. 25). The *torsolike* gene is an exception; mosaics show that its place of action is in the follicle cells that surround the egg. The current model is that *torsolike* is involved in making ligand (by the most anterior and most posterior follicle cells) that binds to the *torso* receptor in those regions and initiates the downstream responses at the poles of the embryo.

The evidence that *torsolike* is required only in the **poles** of the follicular epithelium is convincing. *fragile chorion* is a gene that is required in the follicle cells to make the chorion properly; in its absence the chorion is flimsy. The locus for *fragile chorion* is on the same chromosome arm as *torsolike* so it can act as a cell marker for clones of *torsolike⁻* made by mitotic recombination (Box 4.1, p. 82). So *fragile chorion⁻ torsolike⁻* clones can be made in the follicular epithelium as it grows and produce patches of defective chorion in an otherwise normal egg shell. If these flimsy patches of chorion occur away from the poles the embryo develops normally; if a patch is at the anterior pole, then the embryo lacks the head part, the acron; if it is at the posterior pole then the embryo lacks the tail part, the telson. Clearly, the *torsolike* gene acts in the follicle and it acts locally, being required only in the terminal regions of the follicle to specify the corresponding terminal regions of the embryo.

Figure 2.7 summarises the present model of the sequence of events in the terminal system. Follow the red arrows for the genetic pathway. During oogenesis the 16 cells coming from a germ cell become surrounded by follicle cells that derive from the mesoderm of the mother. Some cells of the follicle leave the surrounding epithelium and migrate inside to the anterior pole of the oocyte; these, as well as two cells at the posterior pole of the egg, have special properties (pink). These cells may release a ligand for the *torso* receptor and this is then stored locally, possibly in the vitelline membrane (red). Later, after the egg is laid, development begins, the *torso* receptor is deployed in the cell membrane and the ligand is released. This ligand is thought to diffuse through the perivitelline fluid that fills the space between the egg plasma membrane and the vitelline membrane and bind to the *torso* receptor at the poles of the egg, leading to a gradient of activation in the embryo.

In summary, the anteroposterior pattern of the newly laid egg depends on three systems that work in different ways. Two (the anterior, *bicoid*, and the posterior, *nanos* systems) rely on the localisation of specific RNA at the poles of the egg, while the third (the terminal, *torso* system) depends on localised effects within the follicle. This picture is summarised diagrammatically in Figure 2.8. In A the three RNAs are shown, in B the protein levels. First, the *nanos* gene product inactivates the *hunchback* RNA in the posterior half of the egg and this leads to a distribution of *hunchback* protein — high anterior, absent posterior and a graded zone in between. The *hunchback* gradient steepens and sharpens as zygotic *hunchback* RNA is translated in the

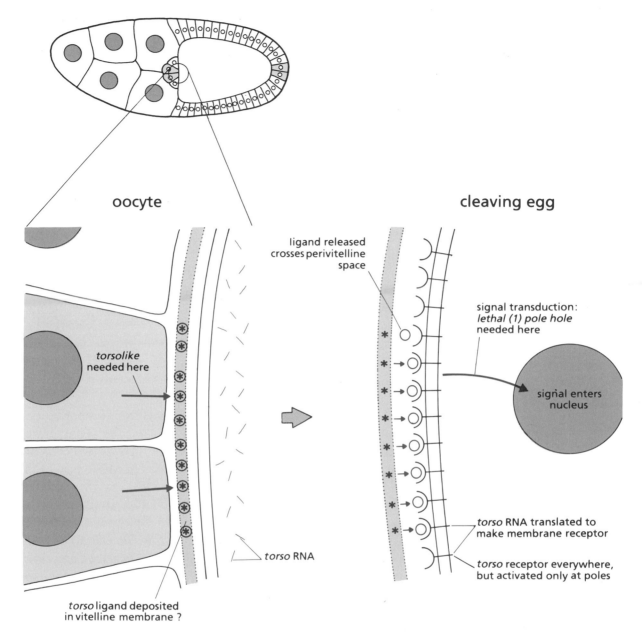

oocyte

cleaving egg

ligand released
crosses perivitelline
space

signal transduction:
lethal (1) pole hole
needed here

torsolike
needed here

signal enters
nucleus

torso RNA

torso RNA translated to
make membrane receptor

torso receptor everywhere,
but activated only at poles

torso ligand deposited
in vitelline membrane ?

Figure 2.7 The terminal system — a model. Local gene expression in the follicle cells results in localised activation of the *torso* receptor in the egg. Only some of the genes in the pathway are known.

anterior region. In C you see graded activation of the *torso* receptor. Double mutants show that these three systems are largely independent and the triple mutant *bicoid⁻ nanos⁻ torso⁻* lacks all pattern — suggesting that there are no other maternal systems needed to lay down the anteroposterior scaffold upon which the embryo is built.

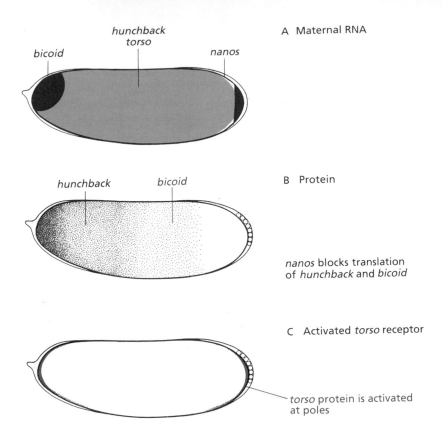

Figure 2.8 The anteroposterior patterning of the egg is determined by three independent systems. RNAs from the *bicoid* and *nanos* genes are localised at the anterior and ? posterior ends of the egg. These are translated into proteins on egg laying. The *torso* receptor is ubiquitous, but is only activated at the poles.

The dorsoventral system

When Sander divided the early egg across and added together the parts of the embryo made by both the anterior and posterior fragments, pieces of pattern were missing. By contrast, dividing the egg in the long axis led to partial duplication of the embryo, and sometimes complete twins (see Figure 2.1). Both the anteroposterior and the dorsoventral system responded differently and independently and he concluded that they were distinct.

The genetic analysis carried out by Nüsslein-Volhard and colleagues has confirmed this, for they have found a family of maternal genes whose wildtype role is to make the dorsoventral pattern of the embryo, but not the anteroposterior. The story is incomplete, but there is an extraordinarily long chain of events that begins in the oocyte, goes out to the surrounding follicle, returns across the perivitelline space and then, entering the cell, results eventually in a protein gradient in the dorsoventral axis of the embryo.

We begin at the beginning with the genes which, as far as we know now, initiate the chain of events. The pathway is summarised in Figure 2.9; start in the oocyte and follow the red arrows. Genes have been

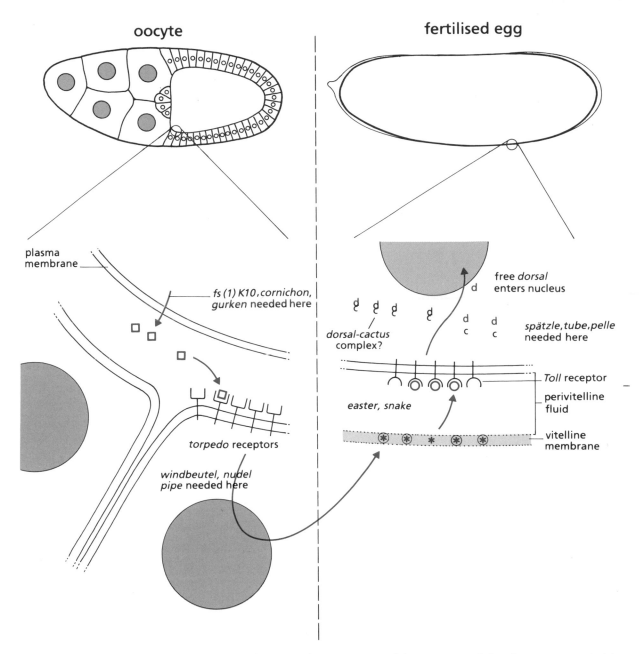

Figure 2.9 The dorsoventral system — a model. It is envisaged that the *cactus* protein (c) binds to the *dorsal* protein (d). In the ventral region of the egg, activation of the *Toll* receptor leads to their dissociation; the free *dorsal* protein can then enter the embryonic nuclei.

identified, for example *cornichon*, *gurken* and *fs(1)K10*, which are required in the **oocyte** for the proper development of the surrounding follicle cells. Mutations in the three genes mentioned above produce odd follicles and embryos that are defective in the dorsoventral axis.

Commonly, both the chorion pattern and embryos are coordinately dorsalised or ventralised; in a dorsalised embryo the cuticle and amino-serosa are extended and the ventral cuticle and mesoderm are reduced or eliminated (Figure 3.8). As with Sander's experimental results and many other maternal-effect phenotypes, the change of pattern is not brought about by the death of some cells but by an abnormal specification of all the cells.

At least one of the functions of the genes needed in the oocyte is probably to send a signal to the follicle, because in the follicle there is a gene (*torpedo*) that encodes a receptor required for the dorsoventral axis of the embryo. The *torpedo* protein is partially homologous to a growth factor receptor (the EGF receptor) of vertebrates and has an extracellular portion that might bind to a ligand from the oocyte. It is not known whether the ligand is **locally** produced in the oocyte, but once the ligand binds to the *torpedo* receptor, its presence will be signalled into the follicle cells. The next step seems to be the production in the follicle of a localised factor that will, somehow, go back to egg. The *nudel*, *pipe* and *windbeutel* genes are required to produce this factor from the follicle and the receptor protein of the *Toll* gene receives the signal in the oocyte.

The *Toll* gene presents an interesting paradox. It is necessary for dorsoventral polarity and in its absence the embryo is strongly dorsalised. It is not only necessary for the embryonic axis but cytoplasmic transfer shows that it can define it. In experiments reminiscent of those done with the *bicoid* gene (p. 29), wildtype cytoplasm was transplanted into a *Toll*$^-$ egg. If this cytoplasm is injected dorsally it can lead to a reversed embryonic axis; the ventral part of the embryo forms in what was, anatomically, the dorsal side of the egg. The paradox is that the *Toll* protein is normally distributed uniformly all over the egg, which would not seem to fit with these local effects. However there is a plausible explanation: it is likely that, in wildtype eggs, the ligand is produced locally and is rapidly bound and mopped up by *Toll* receptors nearby. In *Toll*$^-$ embryos, there is no receptor, so the ligand might be free to diffuse all through the perivitelline space and become evenly distributed. When, following transfer of *Toll*$^+$ cytoplasm, the *Toll* receptor appears locally on the egg membrane, it could be immediately activated at that site to define the ventral region of the embryo.

The local activation of the *Toll* receptor in the fertilised egg is thought to depend on some signal that has been made during oogenesis, deposited somewhere — perhaps on the vitelline membrane — and later transferred across to the most ventral part of the embryo. This transfer probably involves amplification, for other genes have been identified as maternal-effect dorsalising mutations. Two of these genes (*snake* and *easter*) encode secreted serine proteases similar to those found in the amplification cascade of blood clotting and these act within the perivitelline fluid that lies between the egg cell and the vitelline membrane — this has been nicely proven by transferring just perivitelline fluid from *easter*⁺ eggs to the perivitelline space in *easter*⁻ eggs and completely rescuing the mutant.

So we now have an activated receptor along the ventral part of the egg and this is transformed into a gradient of the *dorsal* protein. *dorsal* is a key downstream gene and was singled out amongst the genes with dorsalising mutations by several criteria. As changes in the distribution of *dorsal* protein are made, so changes in the fate map result. Cytoplasmic transfer shows there is more effective *dorsal* product in the ventral part of the egg, and genetic experiments on epistasis (Boxes 2.3 and 2.4) place *dorsal* downstream of all other genes in the pathway

Box 2.4 Molecular epistasis

Now that there are probes, either RNA or antibodies, for gene products, gene order can be determined with the help of these. Consider *fushi tarazu (ftz)* as an example, which is expressed in seven stripes (p. 96). All maternal-effect mutations that are involved in establishing the anteroposterior pattern (such as *bicoid*⁻ (p. 29)) affect the *ftz* stripes, as do mutations in the gap genes, such as *Krüppel*⁻. Even mutants in some other striped genes, such as *hairy*⁻, alter the stripes, so it must be that all these genes act upstream of the *ftz* gene. But mutants in segment polarity genes, such as *wingless*⁻, do not affect *ftz* expression, indeed in *ftz*⁻ embryos **they** are altered. This places the *ftz* gene downstream of some pair rule genes, such as *hairy* and upstream of segment polarity genes, such as *wingless*. It doesn't tell you the mechanism of interaction between the genes, if any: simply the position in some temporal or functional hierarchy. It can be useful, particularly in the negative aspect. If *ftz* expression is completely normal in *engrailed*⁻, that rules out any role of the *engrailed* gene in *ftz* regulation.

There are other ways of finding out the same thing: for example, a dominant gain-of-function mutation can be useful. Take *Toll*ᴰ as an example. *Toll*ᴰ is a constitutively active form of the *Toll* receptor protein (p. 44) which ventralises the embryo. If *Toll*ᴰ is combined with *dorsal*⁻ (which dorsalises the embryo) the result is a dorsalised embryo. One can conclude that *dorsal* is downstream of *Toll*, that *dorsal* is nearer to the final outcome in determining ventral rather than dorsal pattern.

Translocation of gene products between mutant embryos can also help order gene function. For example, if gene *a* is upstream of gene *b*, cytoplasm from *b*⁺ embryos may rescue *a*⁻ embryos, but *a*⁺ cytoplasm will not rescue *b*⁻ embryos. Similarly, translocation of perivitelline fluid helped sort out where in location and where in the hierarchy the gene products of the serine proteases *easter* and *snake* act in the dorsoventral pathway (see above).

Heat shock constructs (Box 3.2, p. 56) can also be useful: flies transformed with a downstream gene under heat shock control will produce phenotypes even in mutants for upstream genes.

shown in Figure 2.9. The *dorsal* protein has been studied and there is a gradient in the dorsoventral axis, almost as expected. The unexpected observation is that the gradient forms from an even distribution of *dorsal* protein, with more protein being taken up into the nuclei along the ventral part and less and less so dorsally. So the gradient is of protein concentration in the nuclei. This is confirmed when the nuclei divide between cleavage cycles 10–13, for, in each of these cycles, the protein is released when the nuclear membranes break down and, temporarily, no gradient is visible. There is some evidence from the sequence homology that *dorsal* protein acts as a transcription factor, which means that the concentration in the nuclei is what counts. In another mutation, *cactus*⁻, with a ventralised mutant phenotype, the *dorsal* protein becomes concentrated in the nuclei all around the dorsoventral axis. One hypothesis is that the *cactus* protein binds to the *dorsal* protein and prevents it entering the nuclei — then, in wildtype embryos, the *cactus* protein is altered in the ventral part of the egg and lets go of the *dorsal* protein — all as the indirect result of the activation of the *Toll* receptor protein.

The present model of this complex chain of events is summarised in Figure 2.9. The end product of all this is a gradient of *dorsal* protein and it seems clear that *dorsal* acts as a morphogen, meaning that the local concentration of *dorsal* (in this case, in the nuclei) determines the pattern of differentiation at each point on the dorsoventral axis. See Figure 3.8 (p. 67) for the distribution of *dorsal* protein in the nuclei of various mutants and the way that distribution correlates with the dorsoventral pattern of the larva.

In this chapter, we have sketched out the way the maternal systems lay down foundations building the fly. As the embryo begins to develop, zygotically encoded genes use these foundations to produce a more elaborate framework. This is explained in Chapter 3.

Further reading

REVIEWS

Anderson, K.V. (1987) Dorsal–ventral embryonic pattern genes of *Drosophila. Trends Genet.* **3**: 91–97.
Gottlieb, E. (1990) Messenger RNA transport and localisation. *Curr. Opin. Cell Biol.* **2**: 1080–1086.
Govind, S. and Steward, R. (1991) Dorsoventral pattern formation in *Drosophila*: signal transduction and nuclear targeting. *Trends Genet.* **7**: 119–125.
King, R.C. (1970) *Ovarian Development in Drosophila melanogaster.* Academic Press, New York.
Levine, M. (1988) Molecular analysis of dorsal–ventral polarity in *Drosophila. Cell* **52**: 785–786.
Nüsslein-Volhard, C. (1991) Determination of the embryonic axes of *Drosophila. Development* (Supplement) **1**: 1–10.

Nüsslein-Volhard, C. and Roth, S. (1989) Axis determination in insect embryos. *Ciba Found. Symp.* **144**: *Cellular Basis of Morphogenesis*, pp. 37–55.

Nüsslein-Volhard, C., Frohnhöfer, H.G. and Lehmann, R. (1987) Determination of antero-posterior polarity in *Drosophila*. *Science* **238**: 1675–1681.

Sander, K. (1975) Pattern specification in the insect embryo. *Ciba Found. Symp.* **29**: *Cell Patterning*, pp. 241–263.

Struhl, G. (1989) Morphogen gradients and the control of body pattern in insect embryos. *Ciba Found. Symp.* **144**: *Cellular Basis of Morphogenesis*, pp. 65–91.

bicoid

Driever, W. and Nüsslein-Volhard, C. (1988a) A gradient of *bicoid* protein in *Drosophila* embryos. *Cell* **54**: 83–93.

Driever, W. and Nüsslein-Volhard, C. (1988b) The *bicoid* protein determines position in the *Drosophila* embryo in a concentration-dependent manner. *Cell* **54**: 95–104.

Frohnhöfer, H.G. and Nüsslein-Volhard, C. (1986) Organization of anterior pattern in the *Drosophila* embryo by the maternal gene *bicoid*. *Nature* **324**: 120–125.

dorsal

Roth, S., Stein, D. and Nüsslein-Volhard, C. (1989) A gradient of nuclear localization of the *dorsal* protein determines dorsoventral pattern in the *Drosophila* embryo. *Cell* **59**: 1189–1202.

Rushlow, C.A., Han, K., Manley, J.L. and Levine, M. (1989) The graded distribution of the *dorsal* morphogen is initiated by selective nuclear transport in *Drosophila*. *Cell* **59**: 1165–1177.

Dorsoventral axis

Manseau, L.J. and Schüpbach, T. (1989) *cappuccino* and *spire*: two unique maternal-effect loci required for both the anteroposterior and dorsoventral patterns of the *Drosophila* embryo. *Genes Dev.* **3**: 1437–1452.

Sander, K. (1971) Pattern formation in longitudinal halves of leaf hopper eggs (Homoptera) and some remarks on the definition of 'embryonic regulation'. *Wilhelm Roux's Archives* **167**: 336–352.

Schüpbach, T. (1987) Germ line and soma cooperate during oogenesis to establish the dorsoventral pattern of egg shell and embryo in *Drosophila melanogaster*. *Cell* **49**: 699–707.

Experimental embryology — ligatures

Herth, W. and Sander, K. (1973) Mode and timing of body pattern formation (regionalization) in the early embryonic development of cyclorrhaphic dipterans (*Protophormia*, *Drosophila*). *Wilhelm Roux's Archives* **172**: 1–27.

Schubiger, G., Moseley, R.C. and Wood, W.J. (1977) Interaction of different egg parts in determination of various body regions in *Drosophila melanogaster*. *Proc. Natl. Acad. Sci. USA* **74**: 2050–2053.

hunchback

Tautz, D. (1988) Regulation of the *Drosophila* segmentation gene *hunchback* by two maternal morphogenetic centres. *Nature* **332**: 281–284.

Tautz, D., Lehmann, R., Schnürch, H., Schuh, R., Seifert, E., Kienlin, A., Jones, K. and Jäckle, H. (1987) Finger protein of novel structure encoded by *hunchback*, a second member of the gap class of *Drosophila* segmentation genes. *Nature* **327**: 383–389.

nanos

Hülskamp, M., Schröder, C., Pfeifle, C., Jäckle, H. and Tautz, D. (1989) Posterior segmentation of the *Drosophila* embryo in the absence of a maternal posterior organizer gene. *Nature* **338**: 629–632.

Irish, V., Lehmann, R. and Akam, M. (1989) The *Drosophila* posterior-group gene *nanos* functions by repressing *hunchback* activity. *Nature* **338**: 646–648.

Lehmann, R. and Nüsslein-Volhard, C. (1991) The maternal gene *nanos* has a central role in posterior pattern formation of the *Drosophila* embryo. *Development* **112**: 679–693.

Sander, K. and Lehmann, R. (1988) *Drosophila* nurse cells produce a posterior signal required for embryonic segmentation and polarity. *Nature* **335**: 68–70.

Struhl, G. (1989) Differing strategies for organizing anterior and posterior body pattern in *Drosophila* embryos. *Nature* **338**: 741–744.

Wang, C. and Lehmann, R. (1991) Nanos is the localised posterior determinant in *Drosophila*. *Cell* **66**: 637–647.

oskar

Lehmann, R. and Nüsslein-Volhard, C. (1986) Abdominal segmentation, pole cell formation, and embryonic polarity require the localized activity of *oskar*, a maternal gene in *Drosophila*. *Cell* **47**: 141–152.

RNA localisation

Frohnhöfer, H.-G. and Nüsslein-Volhard, C. (1987) Maternal genes required for the anterior localization of *bicoid* activity in the embryo of *Drosophila*. *Genes Dev.* **1**: 880–890.

Macdonald, P.M. and Struhl, G. (1988) Cis-acting sequences responsible for anterior localization of *bicoid* mRNA in *Drosophila* embryos. *Nature* **336**: 595–598.

St. Johnston, D., Driever, W., Berleth, T., Richstein, S. and Nüsslein-Volhard, C. (1989) Multiple steps in the localization of *bicoid* RNA to the anterior pole of the *Drosophila* oocyte. *Development* (Supplement) **107**: 13–19.

Segmentation genes

Nüsslein-Volhard, C. and Wieschaus, E. (1980) Mutations affecting segment number and polarity in *Drosophila*. *Nature* **287**: 795–801.

snake, easter

Stein, D., Roth, S., Vogelsang, E. and Nüsslein-Volhard, C. (1991) The polarity of the dorsoventral axis in the *Drosophila* embryo is defined by an extracellular signal. *Cell* **65**: 725–735.

Toll

Anderson, K.V., Bokla, L. and Nüsslein-Volhard, C. (1985a) Establishment of dorsal–ventral polarity in the *Drosophila* embryo: The induction of polarity by the *Toll* gene product. *Cell* **42**: 791–798.

Anderson, K.V., Jürgens, G. and Nüsslein-Volhard, C. (1985b) Establishment of dorsal–ventral polarity in the *Drosophila* embryo: Genetic studies on the role of the *Toll* gene product. *Cell* **42**: 779–789.

Schneider, D.S., Hudson, K.L., Lin, T-Y. and Anderson, K.V. (1991) Dominant and recessive mutations define functional domains of *Toll*, a transmembrane protein required for dorsal–ventral polarity in the *Drosophila* embryo. *Genes Dev.* **5**: 797–807.

torpedo

Price, J.V., Clifford, R.J. and Schüpbach, T. (1989) The maternal ventralizing locus *torpedo* is allelic to *faint little ball*, an embryonic lethal, and encodes the *Drosophila* EGF receptor homolog. *Cell* **56**: 1085–1092.

torso

Casanova, J. and Struhl, G. (1989) Localized surface activity of *torso*, a receptor tyrosine kinase, specifies terminal body pattern in *Drosophila*. *Genes Dev.* **3**: 2025–2038.
Sprenger, F., Stevens, L.M. and Nüsslein-Volhard, C. (1989) The *Drosophila* gene *torso* encodes a putative receptor tyrosine kinase. *Nature* **338**: 478–483.

torsolike

Stevens, L.M., Frohnhöfer, H.G., Klingler, M. and Nüsslein-Volhard, C. (1990) Localized requirement for *torso-like* expression in follicle cells for development of terminal anlagen of the *Drosophila* embryo. *Nature* **346**: 660–663.

SOURCES OF FIGURES

For details, see above.
Figure 2.1 See Herth and Sander (1973), Sander (1971, 1975) and Schubiger *et al.* (1977).
Figure 2.2 See Sander (1975).
Figure 2.3 See Driever and Nüsslein-Volhard (1988a,b) and Frohnhöfer and Nüsslein-Volhard (1986).
Figure 2.4 See Frohnhöfer and Nüsslein-Volhard (1987) and St. Johnston *et al.* (1989).
Figure 2.5 Photograph courtesy of R. Lehmann.
Figure 2.6 See Hülskamp *et al.* (1989), Irish *et al.* (1989) and Strühl (1989).
Figure 2.7 See Nüsslein-Volhard and Roth (1989), Nüsslein-Volhard *et al.* (1987), Sprenger *et al.* (1989) and Stevens *et al.* (1990).
Figure 2.8 See Anderson (1987), Nüsslein-Volhard *et al.* (1987) and Sander (1975).
Figure 2.9 See Anderson *et al.* (1985a,b), Levine (1988), Nüsslein-Volhard and Roth (1989), Price *et al.* (1989) and Stein *et al.* (1991).

3 Patterning the embryo

THE FOUR MATERNAL SYSTEMS that establish positional information in the egg are interpreted by zygotic genes. Each of the gradients or graded zones of activated receptors is read out as several narrower bands of zygotic gene products; thus, orthogonal gradients of a few molecules become elaborated into a tartan pattern of many molecules. Initially, the stripes of the tartan pattern have graded and often overlapping boundaries. Local subsets of these gradients and overlaps act to position each stripe of the final patterning genes.

Elaboration: the first zygotic genes

Four maternal systems provide positional information along the egg axes. First, at both poles of the egg there is activated *torso* receptor; it is most active at the poles and grades in towards the centre of the egg. Second, in the anteroposterior axis there is a single gradient in the concentration of *bicoid* protein. Third, there is a posterior zone that is cleared of maternal *hunchback* protein by the graded activity of the *nanos* gene product. Fourth, along the ventral axis there is a peak of activated *Toll* receptor and the highest level of a concentration gradient of *dorsal* protein in the nuclei.

There appears to be a common strategy that is used again and again in pattern formation. In Chapter 2 we saw how the mother establishes, in the very young embryo, systems of positional information which consist of concentration gradients, or graded zones of activated receptor molecules. It is becoming clear that these maternal gradients are used to locate the transcription of zygotic genes (Box 1.3, p. 8) **all** of which appear to be transcription factors. Each of these genes is activated in only a subpart of the gradient; thus a broad gradient becomes interpreted as a number of narrower bands of gene expression. These genes are not expressed in the mother's germ cells so there is no maternal contribution to phenotype. They have mutant phenotypes which remove rather large chunks of the embryo, giving gaps of several segments or large parts of the terminal regions. For the anteroposterior axis, the genes were recognised as a family by Nüsslein-Volhard and Wieschaus and called gap genes (The segmentation genes, p. 201). The bands of gap gene expression have graded edges which may overlap with adjacent bands. The graded parts of the bands and the overlaps are used to activate other genes even more precisely in narrower stripes, which define the position and polarity of the parasegments.

Eventually this process culminates in the allocation of sets of individual cells to specific developmental fates.

The anteroposterior system — interpreting the *bicoid* gradient

There is a zygotic mutation that produces embryos resembling those produced by *bicoid⁻* mothers; this is *hunchback⁻*. As we have seen (p. 35), *hunchback* protein is produced from both maternal and zygotic RNA; it is made in two bands, a broad anterior one and a narrow posterior one; I will ignore the posterior band here.

The anteriorly located band of *hunchback* protein appears first at late cleavage as a gentle gradient that steepens and intensifies during the blastoderm stage, becoming strongest before gastrulation; at that stage it fills most of the anterior half of the egg, from about 100% to 55% Egg Length. The zygotic contribution to the anterior band is completely dependent on the *bicoid* gene; *bicoid⁻* embryos lack it. The *bicoid* gene is not restricted to controlling *hunchback*; it specifies some parts of the head, parts that are unaffected in embryos that lack all *hunchback* gene product. For this reason, *bicoid* is thought to regulate additional genes that are responsible for anterior pattern.

bicoid, hunchback and the molecular vernier

There is strong molecular evidence for a direct interaction between the *bicoid* protein and the *hunchback* promoter. As this is a paradigm example and probably will be echoed in many other cases to be analysed in the future, we shall dwell on it. The purpose of the molecular interaction is to read the gradient like a **vernier** scale — to transform a gradual change in concentration over a long distance into a sharper response of a gene that occurs at a particular concentration threshold. In this case, a concentration gradient of *bicoid* spread over about 70% of the long axis of the egg is interpreted at about 55% Egg Length by the activation of *hunchback* anterior to that level. Again, the result is not an absolutely sharp cut off, but a gradient. One can estimate the steepness of the *bicoid* gradient as, and when, it functions by looking at the changes in the pattern of *hunchback* expression: as the doses of *bicoid⁺* are increased in the mother, so the zone of *hunchback* expression is extended. For each doubling of the dose of maternal *bicoid⁺* (one to two, two to four) the boundary of *hunchback* expression moves about 10% Egg Length — suggesting that normally the concentration of *bicoid* protein halves over that distance. By the late blastoderm (stage 5(3)) the distance over which *hunchback* protein concentration appears to drop from maximal to undetectable is 5—10% of the egg axis, so it follows that, ultimately, a two-fold difference in concentration of *bicoid* protein is enough to switch *hunchback* from fully on to fully off.

The *bicoid* protein contains a homeodomain (see The discovery of the homeobox, p. 216) and this implies that it is a transcription factor and binds to specific DNA sequences: one might expect it to bind directly to the *hunchback* DNA. One approach to this is to study the interaction, monitoring the effect on a reporter gene, such as *cat*. *cat* encodes an enzyme whose activity can be measured with a coloured dye. The entire promoter region of the *hunchback* gene is joined to the *cat* gene and then either injected into embryos or transfected into insect cells in culture. Whether the *hunchback* promoter responds to the *bicoid* protein can then be determined by comparing the amount of CAT activity in *bicoid*$^+$ as against *bicoid*$^-$ embryos, or by cotransfecting the cells with DNA constructs producing *bicoid* protein. In both cases, the expression of *cat* is massively enhanced by the *bicoid* protein.

Another related method is to define enhancer sequences *in vivo* by

---

Box 3.1 Reporter genes

Often, one may not have an antibody against a gene product and yet wish to know when and where the gene is transcribed. *In situ* hybridisations can be done but these may be laborious and the procedure involves chewing up the cells with proteases — thus spoiling the material and making it difficult, or impossible, to map gene expression at single cell resolution. The reporter gene helps get around this problem and was first developed for *fushi tarazu (ftz)*. Attached to the 5′ end of the coding sequence of *ftz* is a 6.1 kilobase fragment and this and the coding sequence is known to be sufficient to give rescue of *ftz*$^-$ flies — it follows that it might direct the expression of the *ftz* gene in a pattern that is close or identical to the wildtype one. This 6.1 kilobase piece is now linked to the reporter, a *lacZ* gene from *E. coli*, and the whole transformed into flies in a P element vector. The result is an embryo that expresses stripes of β-galactosidase in the same pattern as *ftz*. As β-galactosidase can be stained for, both directly and by using antibodies (p. 32) one can easily study the *ftz* pattern (flies have little β-galactosidase of their own).

There are some possible artefacts which can be problematic. For example the pattern of expression may be influenced by sequences that are near the site of insertion and nothing to do with *ftz* — even the vector sequences can cause tissue-specific expression. The β-galactosidase has a relatively long half-life in flies and therefore may still show long after the gene itself has been switched off; this is a feature that can be useful, particularly with genes expressed in the embryo — the later fate of cells expressing a gene is often of interest (see p. 97).

For more information, see Hiromi, Y., Kuroiwa, A. and Gehring, W.J. (1985) *Cell* **43**: 603–613.

hitching up different amounts of the DNA that flanks the responding gene itself to a **permissive** promoter, which itself does not define the level or pattern of expression. This is then attached to a reporter gene, usually *lacZ* (see Box 3.1), and used to determine which parts of the flanking sequence are required for a particular response. The hybrid gene is transformed into flies and the β-galactosidase reporter protein looked for. Several different constructs with the *hunchback* promoter are illustrated in Figure 3.1. A 747 base pair sequence (shown in the second row of Figure 3.1) gives an almost completely normal expression of *lacZ*. The experiments show that a group of 263 base pairs are themselves indispensable and indeed are sufficient alone to give a good response. Using this construct, the response to changing amounts of *bicoid* protein can be followed *in vivo* and this is shown in Figure 3.2. If the number of doses of *bicoid*⁺ is increased in the mother, the concentration of *bicoid* protein is increased in proportion throughout the embryo and the boundary of *lacZ* expression shifts more posteriorly. Returning to Figure 3.1, you can see that even part of the 263 base pair region is sufficient to respond to *bicoid*; two of these 123 base pair subsequences, if arranged in tandem, are more sensitive than one, and the posterior boundary of expression is shifted. If a smaller fragment of

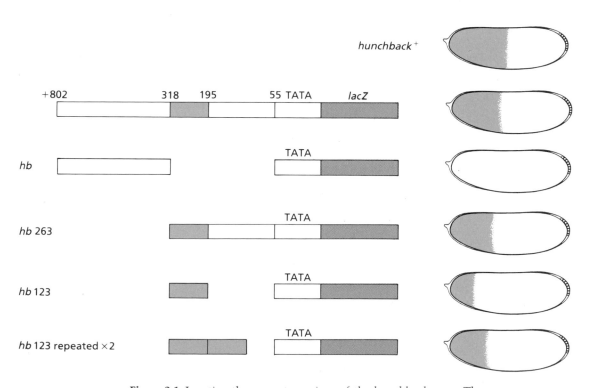

Figure 3.1 Locating the promoter regions of the *hunchback* gene. The constructs illustrated are transferred into flies and are expressed as shown in the eggs.

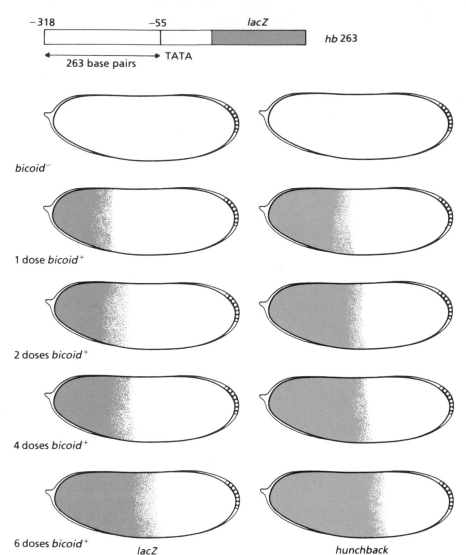

Figure 3.2 The *hunchback* promoter responds to the concentration of *bicoid* protein. A 263 base pair fragment of the *hunchback* promoter will drive expression of the reporter gene, *lacZ* (brown). Transcription is dependent on the amount of *bicoid* protein. As the number of doses of the *bicoid* gene in the mother increases, so the expression of both β-galactosidase (brown) and *hunchback* protein (red) spreads more posteriorly.

82 base pairs is used, there is no response. However, if four copies of this fragment are put end to end, a concentration-dependent response to the *bicoid* protein is found. This and other experiments suggest that cooperative interactions between proteins and the DNA may help increase the sensitivity to the concentration of *bicoid* and the precision of the response. The impression is that the response can be finely and simply adjusted by varying the number and binding affinities of the elements. The following rule applies in some cases: affinity of the site for the binding protein determines the threshold of response, while the number of elements determines the amplitude of the response.

In order to localise where the *bicoid* protein binds to the DNA, footprinting experiments are done. The footprint depends on the DNA to which protein is bound being protected, meaning that it is relatively insensitive to attack by enzymes. Consequently, naked DNA gets cut more easily, more quickly, than the DNA protected by protein. Solutions with and without the *bicoid* protein are bound to DNA which is then partially cut with enzymes *in vitro* and the fragments run side by side on a sequencing gel. The protected sequences can then be seen, because the bands which represent cuts made in the protected parts of the sequence are much weaker or even eliminated in the DNA treated with *bicoid* protein when compared to the DNA that has no *bicoid* protein added. Figure 3.3 shows some DNA sequence that is considerably protected by the *bicoid* protein from the enzyme DNase I. The central

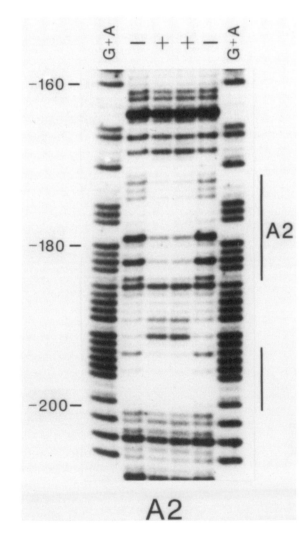

Figure 3.3 The *bicoid* protein associates with a particular part of the *hunchback* promoter, the result is a 'footprint'.

two columns of bands are from DNA digested in the presence of the *bicoid* protein (+) and should be compared with the adjacent ones (−) in which the *bicoid* protein is absent. Note the reduction in band intensity over the area marked A2. The outer columns are used to locate the sequence. Just below the footprint, two bands in the central columns are augmented — such changes are commonplace near footprints and are thought to be due to the bound protein bending the DNA and exposing some sites to increased enzyme attack. The footprint shown, A2, is located in the 123 base pair fragment shown in Figure 3.1. It seems likely, therefore, that the binding site is an important one although it does not work alone. Using this method five binding sites are found in the *hunchback* promoter and they include a common or 'consensus' sequence that is repeated (exactly, or almost exactly) in all five sites. This consensus sequence is TCTAATCCC.

It is no easy matter to find out whether the footprinted regions actually function in the fly in the way envisaged and no such assumption can be made. One probable reason is that the effect of binding a protein such as *bicoid* to a site will vary, depending on whether other protein molecules are also bound nearby and these might well be missing *in vitro*. Further, there may be other proteins which compete with *bicoid in vivo*, or indeed might form a heterodimer with *bicoid*; the combined effect being not activating but repressing. The footprint test is indiscriminate; it may detect binding sites with a meaninglessly low affinity — sites that bind *bicoid* protein can be found even in DNA from the bacteriophage lambda — and some kind of functional assay is required. The best test is to mutate only the sequence of the footprinted

---

Box 3.2 Heat shock promoters

An important part of the tool kit of a *Drosophila* worker is the ability to place genes under the control of unusual promoters. One of the most often used of these is the promoter from the *hsp-70* heat shock gene. The promoter is normally nearly inactive, but can be stimulated to drive expression of the gene assembled at its 3′ end by briefly heating the embryos or flies to 36°C or 37°C for about 30 minutes. This results in universal expression of the gene. For developmental genes, which are normally carefully regulated and are only expressed in particular times and places, this can be a useful trick — the consequences of ectopic expression can tell more about the function of the gene — although it is often lethal.

To make these constructs the *hsp-70* promoter is spliced to the coding region (often from cDNA) of the gene of interest, given a suitable 3′ tail and transformed into flies with a P element vector.

For more information, see Lis, J.T., Simon, J.A. and Sutton, C.A. (1983) *Cell* **35**: 403−410.

site in the promoter, use that to drive a reporter gene in the fly, and to find an altered response to the regulatory gene product; this has not yet been done for the footprint A2. In spite of these technical problems, the *bicoid—hunchback* interaction is a key example of a general phenomenon and it is worthwhile to work out, if possible, exactly how the concentration dependence is achieved.

The *hunchback* gradient and the other gap genes

The gap genes are a group of genes recognised initially by their mutant phenotype: mutant embryos develop with large chunks of the body pattern deleted; the remaining pieces of the pattern may be inappropriately positioned and have reversed polarity. One cannot conclude that the damaged area identifies where the gene acts, because much of the damage can be the **indirect** consequence of the initial malfunction. Apart from *hunchback*, these genes do not show maternal effects (and, in the case of *hunchback*, the maternal contribution is dispensable, see p. 35). Studies of molecular epistasis (see Box 2.4, p. 45) show they are downstream of the maternal coordinate genes (*bicoid, nanos*). The family of gap genes is growing, but the core examples so far studied are *hunchback, Krüppel, knirps, giant, tailless* and *huckebein*. All these genes are expressed in early development, starting during late cleavage and continuing beyond the blastoderm stage. All are expressed in single or double domains in the anteroposterior axis, the domains being fairly broad. All have molecular structures that suggest DNA binding proteins, zinc fingers being the motif in all but the *giant* protein which has another type of DNA binding domain associated with a leucine zipper.

The pattern of gap gene expression is shown in Figure 3.4 where the current estimates of expression are displayed on the anteroposterior axis of the egg. Remember, as the quality of antibodies improves, the apparent widths of the bands tend to increase. Also the patterns are dynamic and the time when they are interpreted, that is when they act, is uncertain.

The wildtype role of the gap genes is understood in outline but many details remain to be worked out. They do two different things; at an early stage they regulate each other, later they define the positions of the pair rule genes. We will consider these in turn.

Gap genes: control and mutual interactions

We have seen that the protein gradient of the *hunchback* gap gene is positioned by the *bicoid* morphogen. In its turn, the *hunchback* gradient itself acts as a morphogen to deploy other gap genes, even though some of the gap genes also respond directly to *bicoid*. The model is that the *hunchback* gradient is read at at least four thresholds: first, above a

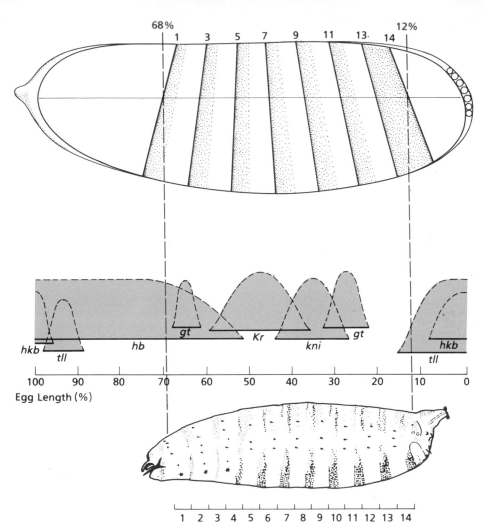

Figure 3.4 The approximate pattern of gap gene expression at the blastoderm (stage 5(2)). At earlier stages, bell-shaped curves and gradients are more broadly spread. The stripes in the embryo indicate the anterior borders of the odd-numbered parasegments and are made by *even-skipped* (Chapter 4).

high value, at about 60% Egg Length, the *Krüppel* gene is repressed, below it the gene is activated — this threshold positions the anterior margin of the *Krüppel* stripe. The posterior margin of the *Krüppel* stripe is fixed by a second and lower threshold, at about 35% Egg Length, below which *Krüppel* is repressed. The third and fourth thresholds, at about 45% and 30%, respectively, position the anterior margins of the *knirps* and the more posterior of the two *giant* stripes — above these thresholds the genes are repressed, below them they are activated. The posterior margins of the *knirps* and *giant* stripes are probably positioned by the *tailless* gradient (p. 64), for in *tailless⁻* embryos the two stripes lose their posterior bounds and the two proteins spread towards the back of the egg. Look at row B in Figure 3.5. In the absence of the *nanos* and *torso* genes, the *hunchback* gradient is abnormal and

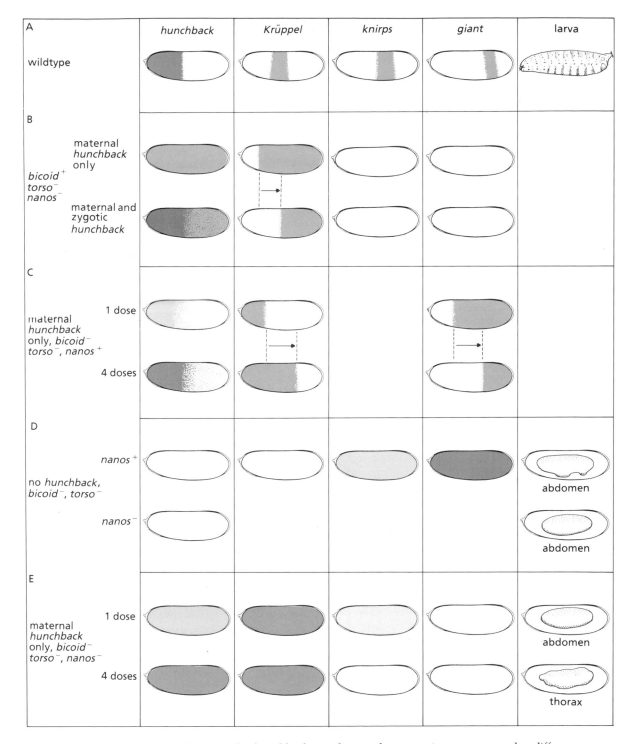

Figure 3.5 The *hunchback* morphogen: these experiments suggest that different concentrations of *hunchback* protein locate the expression *Krüppel*, *knirps* and *giant*. The concentration of the different proteins is indicated by the depth of colour. The shifts induced by increasing *hunchback* protein concentration (arrows) are somewhat exaggerated.

does not reach low levels anywhere. This is true of *hunchback⁻* zygotes (upper row in B), when all the *hunchback* product derives from the mother and the protein is evenly distributed. The small gap at the egg tip where *Krüppel* is not expressed is due to *bicoid* itself or to other gene products dependant on *bicoid*. When the *hunchback* gene is added back to the zygote (lower row in B), the concentration of *hunchback* in the anterior end increases and the anterior edge of the *Krüppel* domain is pushed backwards — evidence that *hunchback* protein is the key agent responsible for repressing transcription of the *Krüppel* gene.

Now look at row C in Figure 3.5. Eggs from mothers which lack the *bicoid* and the *torso* genes have neither the primary anteroposterior gradient nor the terminal system. In *bicoid⁻* eggs, no zygotic *hunchback* is activated, so the amount of *hunchback* protein initially in the embryo depends only on the number of copies of the *hunchback* gene present in the mother. The *hunchback* product will be cleared from the back part of the egg by the posterior system (*nanos*) establishing a *hunchback* gradient illustrated in the first column of Figure 3.5. What is found is that the *bicoid⁻ torso⁻* embryo is divided into two domains; in the anterior part *hunchback* is found and *Krüppel* is induced, in the posterior part *hunchback* is missing and *giant* is induced. If the number of maternal doses of *hunchback* is increased from one to four, the dividing line between the two domains is shifted toward the posterior (Figure 3.5, row C), indicating that the effective agent is indeed the *hunchback* protein concentration. This shift also suggests that the *hunchback* gradient must be read early, when it is shallow — when 1 and 4 doses would put the threshold in different places, as observed.

This hypothesis can be tested more simply by making eggs in which the *hunchback* protein is absent or evenly distributed at different concentrations (Figure 3.5, rows D and E). First, embryos are made that lack all *hunchback* protein; the maternal contribution must be eliminated and this is done by removing *hunchback⁺* from the mother's germ cells (pole cell transplants, see Box 2.2, p. 35). These cells are also *bicoid⁻*, so there will be no zygotic *hunchback*. They are also *torso⁻* so there should be no difference at the poles of the egg. In the complete absence of *hunchback* protein, *giant* is expressed strongly, and *Krüppel* is not activated at all (Figure 3.5, row D).

If there were no polarity in the egg, the outcome in the larva should be an even field of unoriented cuticle. In fact, the larvae have two patches of denticles, the anterior one always being large and the posterior small; they **are** polarised. The polarity is due to the posterior system, represented by the *nanos* gene. As would be expected, in the absence of all three systems (*bicoid⁻ torso⁻ nanos⁻*) the larvae have no detectable polarity (Figure 3.5, row D).

An even and low level of *hunchback* protein is sufficient to activate the *Krüppel* gene but still allow *knirps* expression. A combination that is normally found in the abdomen and leads to the formation of abdominal denticles (Figure 3.5, row E). If the concentration of *hunch-back* is raised further, the *knirps* gene is switched off and, later on, the epidermis forms an even field of thoracic denticles (Figure 3.5, row E'. The difference between abdomen and thorax is controlled by the bithorax complex of genes (see Chapter 5); raising the level of *hunchback* protein can turn abdomen into thorax, a transformation doubtlessly achieved by the direct or indirect effect of *hunchback* on the bithorax complex.

All these experiments are consistent with the model outlined above that the *hunchback* gradient is interpreted during late cleavage and at least at four different levels. Above a certain threshold, *giant* is repressed, above a higher one *knirps* is repressed and two separate thresholds fix the anterior and posterior bounds of the *Krüppel* stripe. The *Krüppel* band is fairly broad, which implies that, at the critical time, the *hunchback* gradient rises only gently between the lower and the upper thresholds. By blastoderm the *hunchback* concentration falls off sharply at about 55% Egg Length which, if it were interpreted then, would bring the bands of expression of the three gap genes closer together than is observed. Also, localised expression of gap genes can already be detected before the blastoderm stage.

The situation with *giant* is complicated because, in the wildtype, there are two separate peaks of expression (see Figure 3.4); it is the posterior peak that is repressed by *hunchback* protein while the anterior peak normally coincides with a high *hunchback* concentration. The two peaks are therefore controlled quite differently. Indeed, the anterior *giant* peak is eliminated in *bicoid⁻* embryos but only slightly altered in *bicoid⁺ hunchback⁻* embryos, suggesting that the anterior peak is independent of *hunchback* and under the direct control of the *bicoid* morphogen. Under this hypothesis, the *bicoid* gradient would be interpreted at at least two levels; a high level would determine *giant* expression, while a lower level would position the *hunchback* gradient.

To return briefly to the *Krüppel* stripe, whose position is mainly determined by the *hunchback* gradient: it is noticeable that the *Krüppel* stripe is flanked by the two *giant* stripes; this might suggest that it is squeezed between them. Indeed, if *giant* is ubiquitously expressed (by placing the gene under heat shock control) the *Krüppel* stripe is severely repressed. I point this out to emphasise that control of gap gene expression is more complicated than implied so far and that a robust system is achieved by several partially redundant mechanisms. There may even be some direct repression of *Krüppel* by high concentrations of *bicoid*. All these layers of control may help stabilise the embryo, but tend to destabilise the mind of the embryologist. Experimental results can bring us back to earth, and the results shown in Figure 3.5 suggest that a fruitful experimental approach is to simplify the system as much as possible, so that interactions can be studied in relative isolation.

Box 3.3 'Blue jumps' or 'enhancer traps'

An ingenious way of using the reporter gene has been developed. If the *lacZ* gene is hitched up to a weak promoter, that construct becomes sensitive to nearby regulative elements — so the pattern of gene expression; the pattern of blue patches after staining for β-galactosidase activity, is determined by enhancers belonging to genes nearby. For example, if the construct inserts into the *engrailed* gene, the β-galactosidase is found in all the cells in the stripes expressing the *engrailed* protein. If the construct is placed in a transposing vector it can be jumped around the genome and stocks showing interesting patterns of blue patches isolated. These stocks are not rare; more than half of the strains give particular patterns of blue staining — cell- and organ-specific enhancers are **very** common in the genome. The stocks can be used themselves — for example to mark particular cell types. They are also useful as a means to clone the nearby gene that directs this pattern of expression. Genomic clones containing the plasmid can be selected, labelled, hybridised to chromosomes, used to isolate homologous cDNAs, perform northerns and start walks. The segment polarity gene *cubitus interruptus* was recently cloned in this way and the blue jumping method is enjoying considerable popularity at the moment.

For more information, see O'Kane, C.J. and Gehring, W. (1987) *Proc. Natl. Acad. Sci.* **84**: 9123–9127. See also Wilson, C., Pearson, R.K., Bellen, H.J., O'Kane, C.J., Grossniklaus, U. and Gehring, W.J. (1990) *Genes Dev.* **3**: 1301–1313.

Not all gap genes are regulated by the anteroposterior system; some have their expression positioned by the terminal genes, a link that should be considered before we progress.

The terminal system

Soon after fertilisation, the *torso* receptor becomes activated at both poles. Although the *torso* receptor itself is distributed evenly, the distribution of **activated** receptor is unknown; it is unlikely to have a sharp border because activation has depended on the diffusion across the perivitelline fluid of some kind of signal. The activated *torso* receptor may form gradients at each pole of the egg. The boundaries of these gradients probably become broader as they are transferred inside by signal transduction, the signal acting on a zygotic gene. The anterior gradient must overlap with the *bicoid* gradient.

There are a number of arguments for *torso* being both activated and read as a gradient. Perhaps the best comes from using a weak dominant mutation that produces some constitutively activated *torso* product (*torso*^{RL3}). The effect of the mutation can be varied by choice, because it is temperature sensitive. The experiment (Figure 3.6) shows that the pattern at the posterior end of the larva depends on the **amount** of activated *torso* receptor. *trunk*$^-$ mothers make eggs with no activated *torso* product and the embryos lack the terminal regions. This can be seen in the anteroposterior fate map as displayed by the *fushi tarazu* stripes, where stripe number 7 is missing. As the activity of the *torso* protein is increased step by step by raising the temperature, the pattern is added back progressively, more central more medial elements appearing first, terminal ones last. This shows that higher and higher amounts of activated *torso* protein specify more and more terminal structures.

Embryos developing from *torso⁻* mothers lack the terminal regions, the acron and telson; by contrast a strong *torso* dominant, that is a gain-of-function mutation, forms an embryo consisting entirely of terminal regions, with no body segments forming at all. By analogy with the anteroposterior system, one would expect a zygotic 'gap' gene or genes to respond to the *torso* gradients. One candidate gene is *tailless*. *tailless⁻* embryos are similar to *torso⁻* ones in that they lack parts of

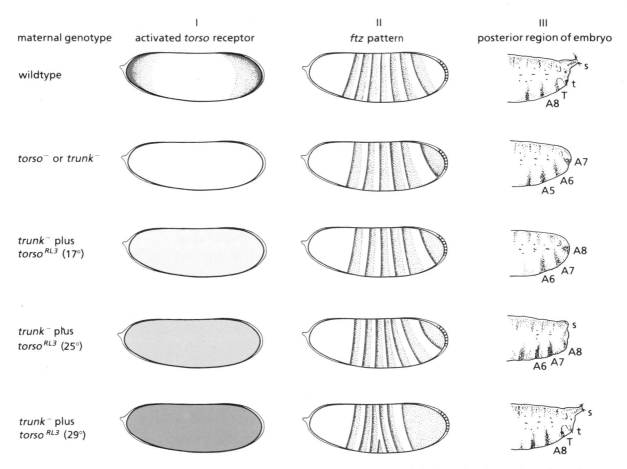

Figure 3.6 The *torso* morphogen. Columns II and III show the observed results, column I is an interpretation (the distribution of **activated** *torso* receptor (red) cannot be seen directly). The dominant mutation *torso^{RL3}* has no effect at 17°C but at 29°C produces nearly enough active receptor to give normal development at the posterior pole of the larva.

the head and parts posterior to A7. These mutants have pieces·missing because allocation of cells has been altered, not because the cells that would have made the missing parts have died. Thus the trunk region of the body expands and the surviving terminal segments are larger.

tailless‾ partially suppresses the *torso*-dominant phenotype in double mutants (showing that the *tailless* gene is downstream of, and dependent on, the *torso* gene, p. 38). If *tailless* were the only gap gene dependent on *torso*, then the phenotypes of *tailless‾* and *torso‾* would be identical but they are not — the phenotype of *tailless‾* is less extreme than *torso‾*. For example, *tailless‾* embryos have some anterior and posterior midgut but *torso‾* embryos do not. Another gap gene, *huckebein*, has been identified and *huckebein‾ tailless‾* double mutants do look like *torso‾* embryos. Moreover, the *torso*-dominant mutant is almost completely suppressed in embryos carrying both *huckebein‾* and *tailless‾*.

The model is quite simple: there are two terminal gradients of activated *torso* receptor, one at each end of the embryo; these are interpreted in a concentration-dependent manner such that, at the highest concentrations at the poles, *huckebein* is activated to specify the anterior and posterior midgut. At a lower concentration, the *tailless* gene is activated. There are some problems; it is not clear how the steps are actually achieved and there is a missing link between the activation of the *torso* receptor and the transcription of the two gap genes. It is probable that activated *torso* receptor phosphorylates a ubiquitous maternal gene product (*lethal (1) polehole*, a homologue of the vertebrate oncogene, *raf*), which is itself a serine/threonine kinase, and this phosphorylates the product of a missing gene, presumably a transcription factor. By analogy with the other systems one might expect this transcription factor, which is as yet unidentified, to be graded and act directly on the downstream genes, *tailless* and *huckebein*.

The *tailless* gene product is a zinc finger protein of the steroid receptor type; it forms a gradient that overlaps anteriorly with the *hunchback* domain and posteriorly with the *giant* domain. Posteriorly it reaches up to the end of the *knirps* gradient. It seems likely that, as with the other gap genes, it acts on even more tightly localised transcription factors downstream.

The dorsoventral system

In Chapter 2 we left this system as a gradient of the *dorsal* protein most concentrated in the nuclei along the ventral midline. It seems likely that the dorsal protein is a transcription factor and, as in other systems, it seems likely that the concentration gradient is interpreted by promoter elements of downstream genes which are expressed exclusively in the zygote. Again, the gradient of a single protein becomes overlapping stripes of several proteins, some also encoding transcription factors of diverse kinds. Information on this system is coming in at a bewildering speed and the story becomes more complicated. One present problem is that, although some of the downstream genes have been found, it is unknown whether these respond directly to *dorsal*, or whether there are more levels of interpretation.

It is probably best to start with what is observed. The *twist* gene product is expressed at late stage 5 in a ventral band of cells that will form the mesoderm. In *twist⁻* embryos the process of gastrulation begins but soon goes wrong. The *twist* protein sequence indicates a helix—loop—helix motif and the protein is localised to nuclei; it is presumably another transcription factor. Initially, the boundaries of *twist* expression are not sharp (Figure 3.7) but, by gastrulation, the borders become more definite and it is the *twist*-expressing cells that roll in. The *twist* gene remains active in the mesoderm cells until they begin to differentiate and so may be required to specify mesoderm. Another mutation, *snail⁻*, has a phenotype related to *twist⁻* but completely fails to gastrulate. The *snail* gene is also expressed in the

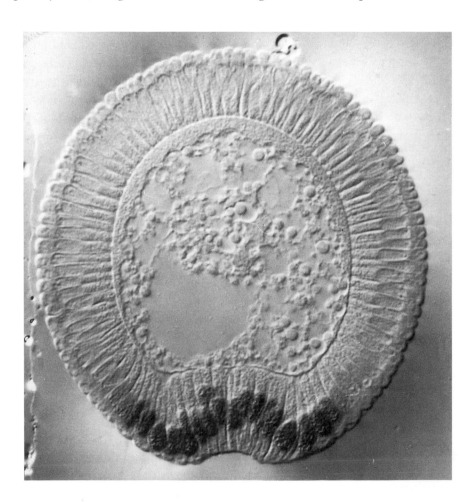

Figure 3.7 Expression of *twist* at the beginning of gastrulation (stage 6), as shown by an antibody against *twist* protein. The ventral nuclei stain strongly.

developing mesoderm; it encodes a zinc finger protein. *snail* may occupy a similar position in the hierarchy to gap genes (which also have zinc finger motifs) in that it is immediately downstream to the morphogen (*dorsal*).

Next to the cells that roll in during gastrulation and express *twist* and *snail* are those that express *single-minded*. The *single-minded* protein is found in nuclei on each side in a single row of cells where mesoderm meets ectoderm — it is later found in the midline neuroblasts (p. 15) and in some glial cells of the central nervous system. The midline neuroblasts and a thin strip of ventral epidermal pattern are defective in *single-minded*⁻ embryos, suggesting that *single-minded* is essential in a subset of cells to specify the identity, the cell type (see Chapter 5). In *snail*⁻ and *twist*⁻ embryos the expression of *single-minded* changes; now cells much nearer the midline contain *single-minded* protein, confirming that the former two mutations cause ventral cells to adopt a more dorsal fate, so that cells that would make mesoderm in the wildtype now make mesectoderm instead.

Another gene that is expressed in a longitudinal band at a characteristic level in the dorsoventral axis is *zerknüllt* (a homeobox gene); the expression pattern is first broad and spread over about 40% of the dorsal blastoderm, but narrows and sharpens to a dorsal strip about five cells wide at stage 6, and this is equivalent to the presumptive serosa. It may be that the gene is involved in specifying the identity of serosa cells, for in *zerknüllt*⁻ embryos the serosa is missing.

The *decapentaplegic* gene is expressed also in the most dorsal blastoderm, at first overlapping with *zerknüllt*, but later becoming most strongly expressed in the cells adjacent to the serosa. *decapentaplegic* is not a transcription factor but instead encodes a protein homologous to the bone morphogenesis proteins of vertebrates (TGFβ family). The protein is secreted and small pieces of about 100 amino acids are chopped off from the C-terminal ends of the molecules. These pieces could diffuse and signal to cells nearby. Unlike nearly all other genes, *decapentaplegic* is haplolethal, which means that 1 dose of the wildtype gene is insufficient to support normal development. This suggests that the exact amount of gene product is important and that *decapentaplegic* could be a morphogen of a kind different from *bicoid*, *hunchback* or *dorsal*. Transcription factors cannot be directly used as morphogens in multicellular systems because they cannot pass from cell to cell; some intermediary mechanism that might involve intercellular signals and receptors may be needed. *decapentaplegic* could be one such signal.

It is clear that these downstream genes are expressed in a pattern that is dependent on the local concentration of *dorsal* protein in the nuclei. As with *bicoid*, if the shape and levels of the *dorsal* gradient are changed by genetic manipulations, there is a corresponding response by the downstream genes, and coordinately of the cell differentiation to mesoderm, nerve cells, ventral and dorsal epidermis. Figure 3.8 gives examples of different distributions of the *dorsal* protein in the nuclei of late cleavage stage embryos and shows the outcome in terms of the distribution of *zerknüllt* and *twist* proteins, as well as the larval pattern. The gradient diagrams are imaginary, but illustrate how different levels in the concentration of *dorsal* protein correspond to different allocations of cell fate. Understanding the downstream genes such as *twist*, *snail*,

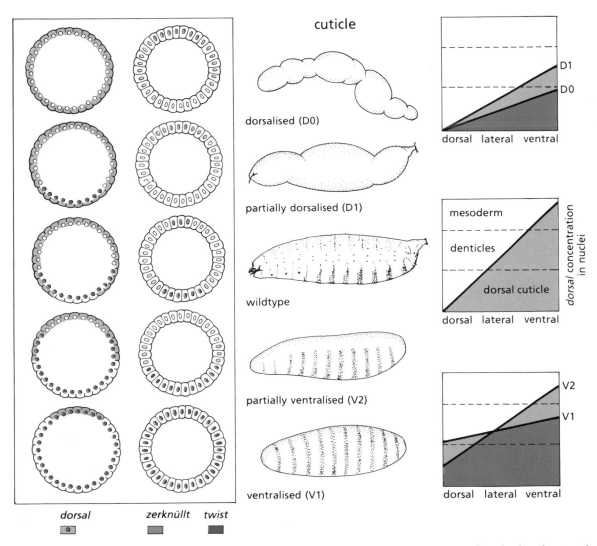

Figure 3.8 The *dorsal* morphogen. Different upstream mutations alter the distribution of the *dorsal* protein which changes the distribution of the downstream genes, *zerknüllt* and *twist* in the blastoderm stage and, later, the larval cuticle pattern. The interpretation offered is shown on the right, where the local concentration of *dorsal* in the nuclei is equivalent to the positional value and determines the fate of the cells. The middle row is wildtype: the diagonal line in the gradient diagrams represents the allocation of cells to mesoderm (most *dorsal* protein in nuclei), denticles, and dorsal cuticle (least *dorsal* in nuclei).

zerknüllt and *decapentaplegic*, finding out how they function and identifying other genes that are involved is a big task for the future.

Elaboration in the anteroposterior axis: the functions of gap genes

In the elaboration that begins with the maternal *bicoid* gradient and ends with the complete map of the embryonic pattern, the gap genes play an intermediary role. There are three separate aspects of pattern formation that are dependent on the gap genes; all three aspects are linked, but it is clear that the gap genes act indirectly on other genes to achieve them. These three aspects are: first, the polarity of the body, polarity that is shown in the proper sequence of body parts — a sequence that can be reversed in the absence of a gap gene. Second, there is the proper spacing of the parasegment primordia; segments can fuse or be widely spaced or missing in gap gene mutants. Third, there is the differentiation of the body parts, for example into thoracic or abdominal patterns, and these are specified in the wrong places in gap gene mutants.

Gap gene products can be removed by mutations or expressed ectopically, that is when the genes are placed under control of a heat shock promoter. Observations on gap gene expression patterns (with antibodies) and on cuticle phenotypes of larvae lead to the idea that the gap genes act together to determine the local pattern. For example, *Krüppel* and *knirps* proteins form offset but overlapping bell-shaped gradients in which, from anterior to posterior, the ratio of *Krüppel* to *knirps* first rises and then falls (see Figure 3.4). These ratios are interpreted coordinately such that the polarity, spacing of parasegments and segmental type are appropriate. If the *Krüppel* gene is removed, we are left with a single symmetric peak of *knirps* which, in a background of high *giant*, is interpreted as a mirror symmetric pattern of **posterior** abdomen segments. When *Krüppel* is expressed universally, the symmetric peak of *knirps* is now formed in an even background of *Krüppel* and a mirror image pattern of **central** abdominal segments results. In the absence of *knirps*, a large area of the posterior embryo is directed by a mix of *Krüppel* and *hunchback* proteins. This combination specifies **anterior** abdomen and thorax; only one or two very large segments may be laid down. Such primordia 'regulate' towards the normal size as excess cells die (p. 152) and the end result is the dwarfish phenotype seen before in *nanos⁻* and *oskar⁻* larvae. These results make sense of the wildtype pattern because they show that the overlapping stripes (for example of *Krüppel* and *knirps*) specify the type of segments that will form and their polarity; it is possible that the bell-shaped curves of concentration made by these stripes may be interpreted as morphogen gradients. They appear to be used to position the next family of genes, the pair rule genes.

The pair rule genes

Mutations in pair rule genes affect every **alternate** segment and the effect is to eliminate a specific part of the cuticle pattern. The major role of pair rule genes is to allocate cells to the 14 parasegments, but they also have minor roles in the head. Two pair rule genes have an immediate function in cell allocation: *fushi tarazu (ftz)* and *even-skipped (eve)*. Larvae mutant for the former (I am told that *fushi tarazu* means 'few segments' in Japanese!) die and lack parts of the abdominal segments of A1, A3, A5, etc., while larvae homozygous for weak alleles of *even-skipped* miss their even-numbered segmental bands. How the mutant phenotypes come about is still mysterious; some, at least, result from death of the cells making part of the pattern. Attention has shifted from that problem to trying to understand the expression patterns of the genes themselves, which is nearer to their wildtype function.

There are a number of pair rule genes, some well studied like *hairy* and *paired*, and others known mainly from their mutant phenotypes. To some extent, they can be arranged in a hierarchy by molecular epistasis (Box 2.4, p. 45). For example *hairy* and *eve* have an early role and have been called primary pair rule genes, meaning that in *hairy*⁻ and *eve*⁻ embryos the secondary pair rule genes such as *ftz* and *paired* show altered patterns of expression (while, in *ftz*⁻ and *paired*⁻ embryos, the *hairy* and *eve* patterns of expression are normal).

The pair rule genes are expressed in stripes in every other segment. Figure 3.9 shows how the *eve* pattern of seven clearly defined stripes develops. Initially, *eve* is expressed at a low level and is seen in all the nuclei; later there is a single broad stripe located anteriorly and this can be detected before the nuclei have finished dividing (stage 4). This stripe narrows and intensifies and becomes stripe number 1, while the others appear more or less together as broad, fuzzy stripes. These gradually narrow, intensify and become asymmetric. As shown in Figure 4.6, p. 98, the anterior margins of the stripes correspond to the anterior boundaries of the odd-numbered parasegments 1–13. Starting a little later, the *ftz* stripes appear and their anterior boundaries eventually delimit the anterior boundaries of the even-numbered parasegments 2–14 (see Chapter 4). The striping pattern is not only there to make pretty pictures, the stripes are crucial to function — so how are they positioned?

The positioning of a stripe

The stripes of *eve* are so regular that they suggest some kind of spatial oscillation, a system of chemical waves. For some time, models of this kind were favoured by the theoreticians. But experiment shows that the stripes form one by one and not as waves. It seems that each stripe

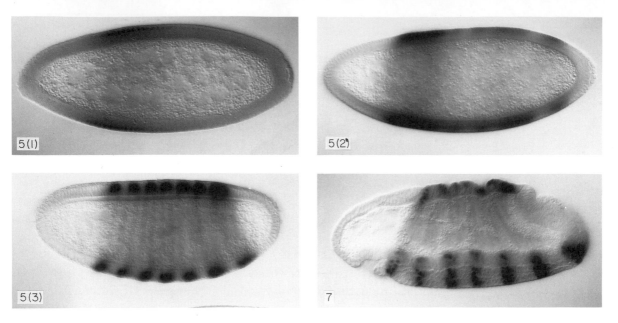

Figure 3.9 The pattern of activation of the *eve* gene in embryos at defined stages (see Figure 1.4).

is directed by largely separate pieces of upstream regulatory DNA and these enhancers can be cut out by restriction enzymes and placed in front of a basal promoter and a reporter gene, *lacZ*, and transformed into flies. When this is done with different small parts of the *hairy* upstream region, the β-galactosidase is synthesised exactly in the position of different stripes. Figure 3.10 illustrates examples in which the three stripes 1, 6 and 7 can be individually made. To demonstrate that the stripes produced really correspond to the endogenous *hairy* stripes, the β-galactosidase and the *hairy* antigen were stained in different colours. Related experiments have been done with *eve* and it is clear that in both genes local regions of the promoter drive individual stripes.

When you have a transgenic fly expressing a single stripe of β-galactosidase, it can then be used to find out which of the upstream genes are responsible for positioning that stripe. This approach has been used with the second *eve* stripe. Why study a construct like this rather than just look at the second *eve* stripe itself? Because it is the only way to **identify** the second stripe in a pattern which may be changed overall with loss, fusion or displacement of stripes. In this case, a properly positioned stripe can be formed if only about 700 base pairs are linked to the *lacZ* gene by a suitable promoter.

The second *eve* stripe is dependent on the *bicoid* gradient as well as the three gap genes *hunchback*, *Krüppel* and the anterior stripe of *giant*: *bicoid* and *hunchback* are permissive; in *hunchback⁻* embryos

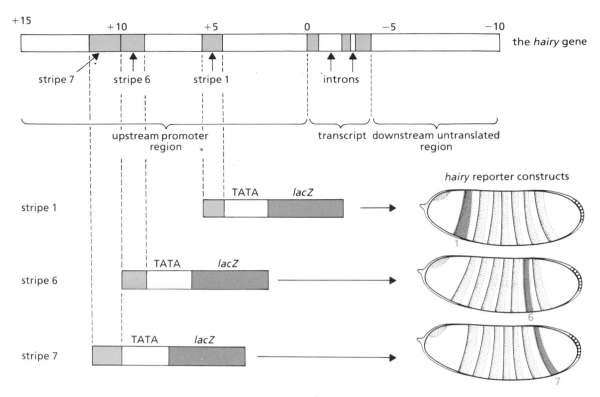

Figure 3.10 The stripes of expression of the *hairy* gene can be formed one by one. Small pieces of the *hairy* promoter are used to drive *lacZ* (brown) in transformed flies.

no stripe develops. Further, in *Krüppel*⁻ embryos the stripe 2 expands posteriorly; in *giant*⁻, stripe 2 expands anteriorly (Figure 3.11C and D). Both these observations make sense because in the wildtype (Figure 3.11A), the *eve* stripe 2 (red) nestles between the twin peaks of *giant* and *Krüppel*. The model is that *eve* can be activated only where there is sufficient *hunchback* protein. Both *Krüppel* and *giant* proteins are repressive — the rising concentration of *giant* on the anterior and the rising concentration of *Krüppel* on the posterior define the limits of the *eve* stripe, so that stripe 2 is squeezed into the valley between the two peaks. Compared with the borders of the two flanking gap genes, which are graded over 5–10 cell diameters, the stripe 2 of *eve* is sharp — it fades from maximum to minimum over only about two cells. This continues with the trend noted before — relatively poorly localised gradients are interpreted to give more precisely localised ones. The process is very sensitive to local concentration. For example, if there is only one dose of the *Krüppel*⁺ gene, the *Krüppel* peak rises up more gently, allowing the *eve* stripe 2 to broaden posteriorly, and to increase in width from five or six to seven or eight cells (Figure 3.11B). There is no certain outcome of this; some embryos go on and

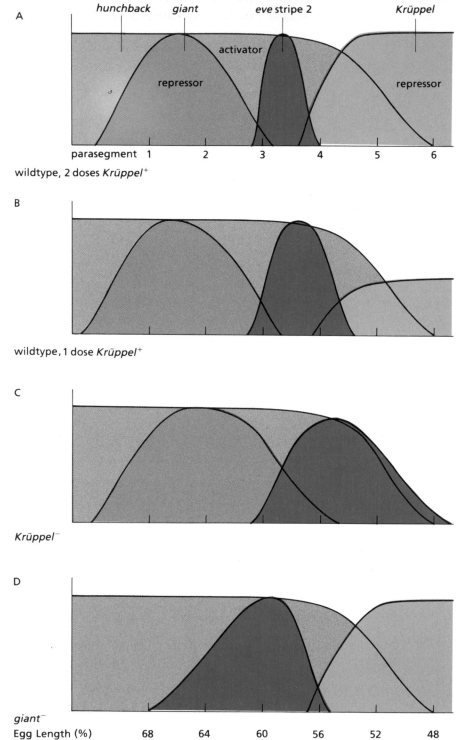

A

hunchback giant *eve* stripe 2 *Krüppel*

activator

repressor repressor

parasegment 1 2 3 4 5 6

wildtype, 2 doses *Krüppel*+

B

wildtype, 1 dose *Krüppel*+

C

Krüppel−

D

giant−

Egg Length (%) 68 64 60 56 52 48

Figure 3.11 The second *eve* stripe (solid red). The *hunchback* (grey) and *bicoid* proteins are permissive. *giant* (brown) and *Krüppel* (pink) proteins are repressive.

give normal flies — developmental systems are impressively robust; minor variations in earlier steps are often corrected later on.

Notice that the anterior and posterior limits of *eve* stripe 2 can be viewed as being fixed independently by exactly the same kind of molecular vernier used to switch on *hunchback* above a particular threshold. For the anterior edge, it is the graded concentration of *giant* that is important, while for the posterior edge, it is the concentration gradient of *Krüppel*.

The next experiment is to take the 700 base pairs of DNA which are necessary to position the stripe 2 correctly, treat the DNA with *bicoid*, *hunchback*, *giant* and *Krüppel* proteins and look for footprints (p. 55). Binding sites for all the proteins are found; they are arranged in clusters and, frequently, activating sites (*bicoid*, *hunchback*) overlap with repressing sites (*Krüppel*, *giant*). In Figure 3.12 the activating sites for *bicoid* and *hunchback* are shown in black and grey and the sites for repressing proteins *giant* and *Krüppel* in brown and pink, respectively. The sites are often adjacent or overlapping; this gives scope for protein–protein interactions. The general idea is that protein interactions serve to sharpen up the response to differing protein concentrations so that small changes in concentration can lead to an on/off switch of *eve* transcription.

One can get some measure of the value of specific binding sites as switches by assays carried out in *Drosophila* cells in culture. The binding sites to be tested are linked to a *cat* reporter gene via a

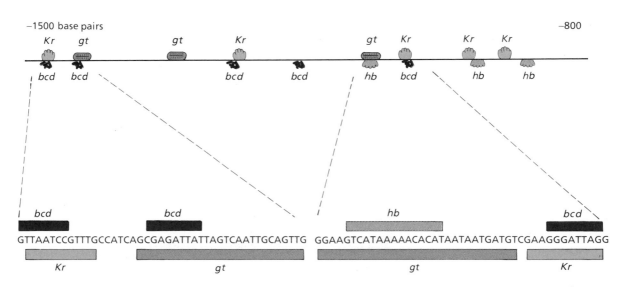

Figure 3.12 *eve* stripe number 2 — the controlling element in the promoter and its binding sites. The sites were mapped by footprints made by the complete proteins transcribed from the *Krüppel* (*Kr*), *giant* (*gt*), *hunchback* (*hb*) and *bicoid* (*bcd*) genes in *E. coli*.

promoter (p. 52). Plasmids containing this construct are transfected into the cultured cells together with different doses of other plasmids expressing the proteins that bind to the sites (e.g. *bicoid*). Using promoter sequences that drive the second *eve* stripe, experiments of this type have shown that two or more activating proteins are far better than one, suggesting that there may be cooperative effects. When plasmids that express both *hunchback* and *bicoid* proteins are placed in the same cells, the CAT level is enhanced much more than with either protein alone. But when *Krüppel* protein is added to an effective mix of *hunchback* and *bicoid*, it quashes the CAT activity, exactly as envisaged for the second *eve* stripe in the embryo. The *giant* protein is an even more effective repressor of *cat* transcription than *Krüppel*, and this is expected because it has to work, *in vivo*, anterior to the *eve* second stripe, where the activating proteins *bicoid* and *hunchback* are at higher concentrations. The mechanism of repression is likely to be competition for the same sites, or for different sites that are so close that there is not room for the two counteracting proteins to bind at once. It is thought that, in the presence of sufficient concentrations of

Box 3.4 Multipurpose genes

The geneticists' and embryologists' approach to a gene has traditionally had one common feature — both are looking for a main function of the gene and have, by and large, disregarded the possibility that a gene could have several distinct roles. Molecular biology has suggested otherwise. For example *Krüppel* and *ftz* have their main early roles in defining zones and parasegment boundaries in the young embryo. However antibodies against these two proteins suggest later functions; they become expressed in specific subsets of neurones and, in *ftz⁻* mutants lacking later functions, these particular neurones are defective. The control elements are special, suggesting that evolution has tacked on extra jobs: of course, it is difficult to be sure which is the oldest and primary function. As more and more genes are cloned, their expression pattern and flanking control regions analysed, this picture of multipurpose genes is widening. Transcription factors that are used for specific early developmental events seem to turn up later in the central nervous system, presumably to contribute to the diversity of cell identities there. Likewise selector genes, such as *Ubx*, have garnered more and more roles as insects have evolved and this makes it very hard to see simple patterns and make useful generalisations. For genes whose function is more mysterious, like the secreted protein *wingless*, it is confusing to see complex expression patterns, both in time and space. The gene only encodes one protein but its purposes could be widely divergent, as may well be the case when it is activated as a segment polarity gene in the early embryo (p. 101), in an inductive role in the visceral mesoderm of the gut later on or along the dorsoventral compartment boundary in the wing imaginal disc, even later.

giant and/or *Krüppel* proteins, *bicoid* and/or *hunchback* proteins might be excluded from the sites.

A further test is to delete or alter only selected binding sites, hitch the otherwise undamaged promoter up to *lacZ* and see the effects on the stripe *in vivo*. Again the results confirm the significance of these sites; for example, deletion of three *giant* binding sites results in anterior broadening of the stripe — exactly as expected from the model.

Although only some details are known, the model in the mind incorporates aspects from other eukaryotic promoters (different proteins binding to the same or overlapping sites, producing activation or repression, competition and cooperative interactions between the proteins; all leading to the switching on or off of genes).

Together, all these experiments give a picture of activation of one stripe and they are sufficiently strong to make one feel hopeful that the process is understood in principle. While most of the results fit the 'central dogma' of the fly embryo (maternal genes define axes and regulate zygotic gap genes that define domains; gap gene proteins control pair rule genes that allocate cells), the evidence for a direct role for the *bicoid* protein on the *eve* stripe 2 promoter, as well as the important interactions between gap genes, indicate many exceptions to this simple linear hierarchy (see also Box 3.4).

The process we have described generates a symmetric stripe of *eve* protein by the middle of the blastoderm period (stage 5(2), see Plate 4.1), but I believe the stripe is not yet ready to act and to allocate cells. That probably occurs when the stripe sharpens anteriorly, a step discussed in the next chapter.

Further reading

REVIEWS

See Further reading, Chapters 1 and 2.

Ferguson, E.L. and Anderson, K.V. (1991) *Curr. Top. Dev. Biol.* In press.

Gaul, U. and Jäckle, H. (1990) Role of gap genes in early *Drosophila* development. *Adv. Genet.* **27**: 239–275.

Hülskamp, M. and Tautz, D. (1991) Gap genes and gradients — the logic behind the gaps. *Bioessays* **13**: 261–269.

Rushlow, C. and Levine, M. (1990) Role of the *zerknüllt* gene in dorsal–ventral pattern formation in *Drosophila. Adv. Genet.* **27**: 277–307.

SELECTED PAPERS

bicoid/hunchback

Driever, W. and Nüsslein-Volhard, C. (1989) The *bicoid* protein is a positive regulator of *hunchback* transcription in the early *Drosophila* embryo. *Nature* **337**: 138–143.

Struhl, G., Struhl, K. and Macdonald, P.M. (1989) The gradient morphogen *bicoid* is a concentration-dependent transcriptional activator. *Cell* **57**: 1259–1273.

eve and hairy stripes

Howard, K.R. and Struhl, G. (1990) Decoding positional information: regulation of the pair-rule gene *hairy*. *Development* **110**: 1223–1231.

Macdonald, P.M., Ingham, P. and Struhl, G. (1986) Isolation, structure, and expression of *even-skipped*: a second pair-rule gene of *Drosophila* containing a homeobox. *Cell* **47**: 721–734.

Pankratz, M.J., Seifert, E., Gerwin, N., Billi, B., Nauber, U. and Jäckle, H. (1990) Gradients of *Krüppel* and *knirps* gene products direct pair-rule gene stripe patterning in the posterior region of the *Drosophila* embryo. *Cell* **61**: 309–317.

Small, S., Kraut, R., Hoey, T., Warrior, R. and Levine, M. (1991) Transcriptional regulation of a pair-rule stripe in *Drosophila*. *Genes Dev.* **5**: 827–839.

Stanojevic, D., Hoey, T. and Levine, M. (1989) Sequence-specific DNA-binding activities of the gap proteins encoded by *hunchback* and *Krüppel* in *Drosophila*. *Nature* **341**: 331–335.

Stanojevic, D., Small, D. and Levine, M. (1991) Overlapping gradients of transcriptional activators and repressors specify a pair-rule stripe in the *Drosophila* embryo. *Science*. In press.

Warrior, R. and Levine, M. (1990) Dose-dependent regulation of pair-rule stripes by gap proteins and the initiation of segment polarity. *Development* **110**: 759–767.

giant

Capovilla, M., Eldon, E.D. and Pirrotta, V. (1991) The *giant* gene of *Drosophila* encodes a b-ZIP DNA-binding protein that regulates the expression of other segmentation gap genes. *Development* **114**: 99–112.

Kraut, R. and Levine, M. (1991) Mutually repressive interactions between the gap genes *giant* and *Krüppel* define middle body regions of the *Drosophila* embryo. *Development* **111**: 611–621.

huckebein

Weigel, D., Jürgens, G., Klingler, M. and Jäckle, H. (1990) Two gap genes mediate maternal terminal pattern information in *Drosophila*. *Science* **248**: 495–498.

hunchback, Krüppel, knirps

Gaul, U. and Jäckle, H. (1989) Analysis of maternal effect mutant combinations elucidates regulation and function of the overlap of *hunchback* and *Krüppel* gene expression in the *Drosophila* blastoderm embryo. *Development* **107**: 651–662.

Hülskamp, M., Pfeifle, C. and Tautz, D. (1990) A morphogenetic gradient of *hunchback* protein organises the expression of the gap genes *Krüppel* and *knirps* in the early *Drosophila* embryo. *Nature* **346**: 577–580.

Pankratz, M.J., Hoch, M., Seifert, E. and Jäckle, H. (1989) *Krüppel* requirement for *knirps* enhancement reflects overlapping gap gene activities in the *Drosophila* embryo. *Nature* **341**: 337–340.

Struhl, G., Johnston, P. and Lawrence P.A. (1992) The *hunchback* gradient specifies body pattern in the *Drosophila* embryo. *Cell* **69**: 237–249.

tailless

Pignoni, F., Baldarelli, R.M., Steingrimsson, E., Diaz, R.J., Patapoutlan, A., Merriam, J.R. and Lengyel, J.A. (1990) The *Drosophila* gene *tailless* is expressed at the embryonic termini and is a member of the steroid receptor superfamily. *Cell* **62**: 151–163.

Zygotic dorsoventral genes (dorsal, twist, single-minded, zerknüllt, decapentaplegic, snail)

Boulay, J.L., Dennefeld, C. and Alberga, A. (1987) The *Drosophila* developmental gene *snail* encodes a protein with nucleic acid binding fingers. *Nature* **330**: 395–398.

Ip, Y.T., Kraut, R., Levine, M. and Rushlow, C.A. (1991) The *dorsal* morphogen is a sequence-specific DNA-binding protein that interacts with a long-range repression element in *Drosophila. Cell* **64**: 439–446.

Leptin, M. (1991) *twist* and *snail* as positive and negative regulators during *Drosophila* mesoderm development. *Genes Dev.* **5**: 1568–1576.

Nambu, J.R., Franks, R.G., Hu, S. and Crews, S.T. (1990) The *single-minded* gene of *Drosophila* is required for the expression of genes important for the development of CNS midline cells. *Cell* **63**: 63–75.

Posakony, L.G., Raftery, L.A. and Gelbart, W.M. (1991) Wing formation in *Drosophila melanogaster* requires *decapentaplegic* gene function along the anterior–posterior compartment boundary. *Mech. Ageing Dev.* **33**: 69–82.

Rushlow, C., Frasch, M., Doyle, H. and Levine, M. (1987) Maternal regulation of *zerknüllt*: a homoeobox gene controlling differentiation of dorsal tissues in *Drosophila. Nature* **330**: 583–586.

Thisse, B., Stoetzel, C., Gorostiza-Thisse, C. and Perrin-Schmitt, F. (1988) Sequence of the *twist* gene and nuclear localization of its protein in endomesodermal cells of early *Drosophila* embryos. *EMBO J.* **7**: 2175–2183.

SOURCES OF FIGURES

For details, see above.

Figure 3.1   After Struhl *et al.* (1989).

Figure 3.2   After Struhl *et al.* (1989). See Driever and Nüsslein-Volhard (1989).

Figure 3.3   From Driever and Nüsslein-Volhard (1989).

Figure 3.4   See Gaul and Jäckle (1989, 1990), Hülskamp *et al.* (1990), Kraut and Levine (1991), Pankratz *et al.* (1989), Pignoni *et al.* (1990) and Weigel *et al.* (1990).

Figure 3.5   After Struhl *et al.* (1992). See Hülskamp and Tautz (1991) and Hülskamp *et al.* (1990).

Figure 3.6   After Casanova and Struhl (1989) (details p. 49).

Figure 3.7   See Thisse *et al.* (1988). Photograph courtesy of M. Leptin.

Figure 3.8   After Anderson. *et al.* (1985a) (details p. 48) and Roth *et al.* (1989) (details p. 47).

Figure 3.9   See Macdonald *et al.* (1986) and Frasch and Levine (1987) (details p. 105).

Figure 3.10 After Howard and Struhl (1990). See Pankratz *et al.* (1990).

Figure 3.11 After Small *et al.* (1991).

Figure 3.12 After Small *et al.* (1991).

4 Cell lineage and cell allocation

*E*XPERIMENTS ON CELL LINEAGE *show that each cell of the ectoderm and (probably) mesoderm is unequivocally allocated to a segmental unit, a parasegment. This is achieved by the pair rule genes, particularly* fushi tarazu *and* even-skipped, *which together allocate the cells to parasegments.* engrailed *subdivides the ectoderm of each parasegment into posterior and anterior compartments.*

Further elaboration: the allocation of individual cells

Before we can go beyond the stripes of pair rule gene expression and see how they allocate cells, we have to digress and describe the states to which cells are allocated. And this means going into older experiments of a different kind, which asked the question, 'how do the cells of the embryo generate the cells of the adult?' *A priori*, there are many possible answers to this question all of which must lie between two extremes. At one extreme there would be no defined pattern of cell descent; cells would just divide and create a jumble of more cells and these cells would be used in clumps to make the adult organs with no regard to their ancestry. At the other extreme, the entire cell lineage could be invariant; cells could divide in a fixed and programmed pattern to generate a lineage tree and organs could arise predictably and rigidly from specific sets of related cells.

This latter extreme is (almost) exactly what happens in the small nematode *Caenorhabditis elegans*; the entire cell lineage of the wildtype worm can be, and has been, written down in the form of a branching tree. However, the nematode is exceptional; the cell lineage of other animals, including flies and humans, is much more indeterminate. Thompson thought that animals are made seamlessly; he wrote 'the living body is one integral and indivisible whole in which we cannot find, when we come to look for it, any strict dividing line even between the head and the body, the muscle and the tendon, the sinew and the bone.' [6] To find out how far this is true for flies we need an objective method that will put an indelible mark on an embryonic cell, a mark that will be displayed by all its descendants in the adult. In *Drosophila*, such marking methods are available and have shown Thompson's statement is not true of development — even though it may be true of anatomy. Insect bodies are made **piecemeal**. The pieces, which consist of many cells, meet at boundaries that are fixed in position. These domains are so well integrated that the boundaries cannot be seen without special methods — there is some resemblance to a jigsaw puzzle, in that the boundaries of the interdigitating pieces may not correspond to features in the picture printed on them.

In *Drosophila*, the methods used to mark cells are genetic (see Boxes 4.1 and 4.2). If a single cell is labelled early in development it will generate a clone of marked cells. If this clone colonises two organs it is clear that, at the time of marking, the cell cannot have been 'determined' between the two organs; it cannot have belonged to a primordium of one organ rather than the other. However, once the founder cells of an organ have been established, then clones descending from a founder cell will be confined to that organ. If the organ exists as a precise population of cells with a defined boundary, then the clone of cells may run up to and along that boundary.

If single cells are marked early during cleavage (see gynandromorphs, p. 11) they can colonise several different organs and germ layers. These large clones prove that cleaving nuclei are undetermined, meaning that their developmental fate is not restricted by any intrinsic constraint. Cells do not divide during the blastoderm stage, so the next moment cells can be genetically marked is at the first division after blastoderm (see p. 11). It so happens that in *Drosophila*, most larval cells do not divide much if at all, so a clone that is induced at that time marks only a few larval cells. This is true of the larval ectoderm, mesoderm and endoderm. For this reason, to find limitations in developmental potential, one must look at adult organs — for these derive from embryonic cells that have divided many times. What is found is surprising and illuminating and is summarised in Figure 4.1. In A, a nucleus is marked during cleavage and its descendants can colonise large and variable regions of the adult. In B, a nucleus is hit by X-rays during the blastoderm stage. When the cell divides with the next mitosis, which occurs in stage 9, a marked daughter cell is formed in the epidermis. That cell, even if it grows relatively rapidly (is *Minute*[+], see Box 4.1), only colonises one precisely defined part of each segment, either the anterior or the posterior part, but **never** both. It follows that cells must have been allocated in the embryo to make either an anterior or a posterior part of each segment. These parts are called 'compartments'. The experiment illustrated in Figure 4.1 also shows that the founding cells for compartments are allocated some time after the beginning of blastoderm and before the epidermal cells enter a second post-blastoderm mitosis.

The experiment also shows that cells marked at blastoderm can generate clones in parts of both the wing and the second (T2) leg, although they are confined, in both appendages, either to the anterior or to the posterior compartment. When clones are produced later in the

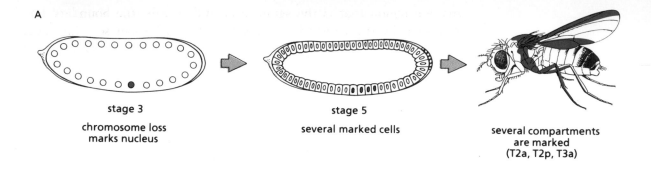

A

stage 3

chromosome loss
marks nucleus

stage 5

several marked cells

several compartments
are marked
(T2a, T2p, T3a)

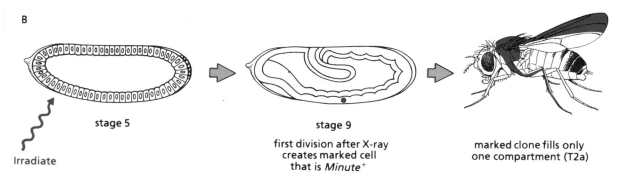

B

Irradiate

stage 5

stage 9

first division after X-ray
creates marked cell
that is *Minute*⁺

marked clone fills only
one compartment (T2a)

Figure 4.1 Early induced clones cross compartment boundaries, later ones do not.

embryo they are confined to either wing or leg. Two compartments have become four compartments. Much later, the wing primordium itself becomes subdivided into two groups of cells that will make dorsal and ventral compartments which meet precisely at the perimeter of the adult wing. It follows that compartments can form progressively — although the mechanisms used in each successive step are probably not the same. These results also establish that in the young embryo the wing and the leg of the adult have a common origin; most likely, there is a combined primordium in the ventral region of the young embryo and, later, the wing cells migrate dorsally.

Using similar methods, but with different cell markers, it has been shown that clones do not cross between the mesoderm and the ectoderm of the same segment, and, within the mesoderm, are probably confined to sets, that is compartments, of muscles (p. 93).

A compartment is an area of the developing or mature fly that is constructed by **all** the descendants of a founding set of cells. When the smaller clones are studied it becomes clear that the way a compartment is made varies from individual to individual (the clones differ in position, size and shape), but together the founding group of clones, the 'polyclone', make a region that is the same in all individuals (the boundary line respected by the clones is always the same). We can see this most clearly in the wing. In Figure 4.2, the top four wings show examples of conventionally marked clones made by irradiating embryos and young

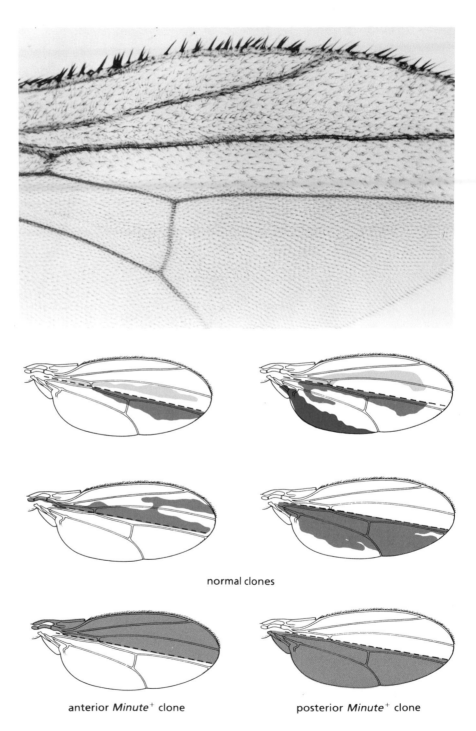

normal clones

anterior *Minute*⁺ clone   posterior *Minute*⁺ clone

Figure 4.2 The construction of compartments in the wing. Red and pink clones are normal and gratuitously marked. Brown clones are marked and *Minute*⁺. Photograph shows a *Minute*⁺ clone in the anterior compartment. It respects the compartment boundary precisely.

larvae (see *engrailed*: the life history of a gene, p. 207). None of the clones crosses the anteroposterior compartment boundary but they vary in size and shape. In spite of all this variation **within** the group of cells that make the wing compartment, **together** they always construct exactly the same portion of the wing. Large *Minute*$^+$ clones make the compartments unmissable, they often fill the entire compartment, colonising both dorsal and ventral wing surfaces (brown clones, Figure 4.2). The photograph in Figure 4.2 shows part of a wing with an anterior *Minute*$^+$ clone filling the anterior compartment.

The cell lineage studies lead to the conclusion that the embryo, before the second division following the blastoderm stage, is divided into stripes of determined cells. These defined groups of cells will generate the anterior or posterior compartments of the adult and, almost certainly, the anterior and posterior compartments of the larva. It is important to emphasise that, for the progenitors of at least the adult epidermis, the lineage studies tell us that **all** the cells are allocated; there are no sets of intermediate cells whose descendants can colonise different compartments.

Box 4.1 Genetic mosaics — mitotic recombination

Embryologists have always wanted to know the cell lineage of an organ. For example, do all the cells that make the heart descend from one or more founding cells? As the heart grows, does each cell divide a similar number of times at fixed intervals? Does the heart share progenitor cells with other organs and, if so, which? Questions like these can be approached by making marked clones of cells in the developing embryo, and the best method is mitotic recombination.

The principle of mitotic recombination is that X-rays induce chromosome breaks that result in exchange of parts between homologous chromatids. This occurs only rarely so that a picture of cell lineage is usually built up piecemeal by looking at clones found in a large number of individuals. Take the wing as an example; since it is made by epidermal cells, a cuticle marker gene is needed. There is a mutation *multiple wing hairs* (*mwh*$^-$); it is viable and fertile but has, on the wing and elsewhere, distinct cuticular hairs. We cross *mwh*$^-$ with *mwh*$^+$ flies and, as the F1 *mwh*$^+$/*mwh*$^-$ embryos of a particular age group grow up, they are irradiated *en masse* with 1000 rads and the surviving adults searched for clones of *mwh*$^-$ cells (which are seen as patches of cells with extra hairs, Figure B4.1).

A great deal of information can be gained by studying clones. The later the irradiation, the higher the frequency and the smaller the size of the clones (because the number of cells in the primordium of the wing, the target size, is increasing) so we can work out the average frequency of divisions and the mean growth rate of the organ. By labelling sister cells ('twin spots' see Figure B4.1B) we can compare the number of divisions followed by sister cells. Does the clone include both cuticle surface and bristles? If it does it must be that single cells are still able to give rise to

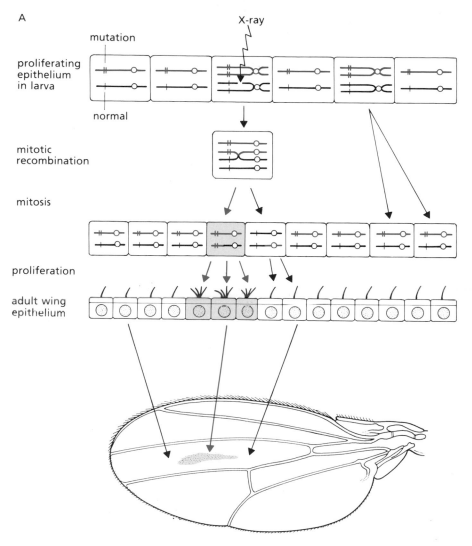

Figure B4.1 Mitotic recombination — the principle and some modifications. A, the principle. In an embryo heterozygous for a mutant that alters the cuticle there will be a paternally and a maternally derived chromosome, one carrying the mutation and one not. X-irradiation causes·breaks and can result in exchange between the two chromatids as shown. After mitosis and chromosome segregation a cell can be formed that is homozygous for the mutation, this cell will found a clone of marked cells that will make a marked patch on the adult. Normally, each cell in the wing secretes a single hair, but in the mutant *bushy* (*bsh*) each cell forms many hairs.

B (*over*), modifications. (i) Twin spot: If there are two marker mutations arranged as shown, irradiation can give two sister clones, one from each daughter cell. (ii) *Minute* technique: Coupling the marker mutant with a *Minute*⁺ allele as shown can generate a marked clone that is wildtype, having lost the deleterious *Minute*⁻ allele. This clone will grow at the wildtype rate in a slow growing host. *Minute*⁺ clones often fill, or nearly fill, compartments. (iii) Dominant female sterile: females carrying an allele of some dominant female steriles do not make eggs. Mitotic recombination can cause the loss of the allele and the establishment of a competent cell in the female germ line. This cell produces a batch of eggs, all belonging to a clone and if the mutant of interest (*m*⁻) is arranged as shown, the eggs will be homozygous for that mutant. (iv) Intragenic recombination: if two differently located mutations in the *white* gene are arranged *in trans* the eye is white but mitotic recombination **between** the two mutations can regenerate the wildtype *white*⁺ gene that will make pigment. Such events are rare but can be useful (see Chapter 8 and Figure 8.4 and Plate 8.1).

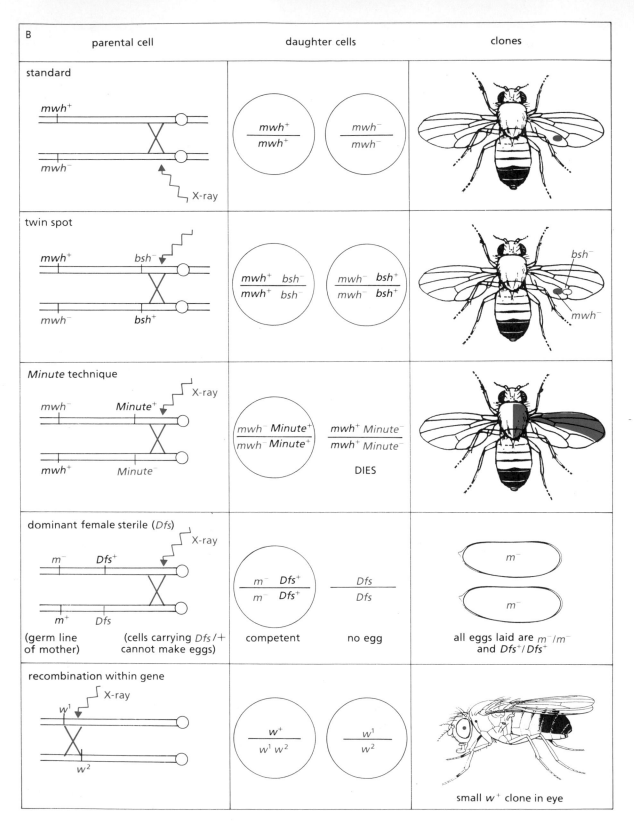

Figure B4.1 (*continued*)

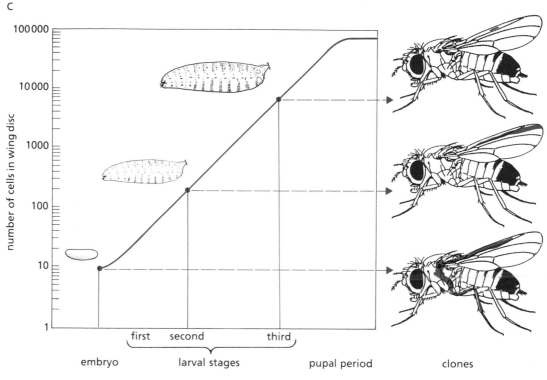

C

Figure B4.1 (*continued*) C, when clones are induced early they make larger patches on the adult than when they are produced late.

both cell types. Later in the development, clones may mark either bristles or wing surface but not both, showing that separate sets of cells now make these structures. Remember, in these experiments, cells are marked one division after the irradiation, because the mutant cell is only formed after chromosome segregation.

Is there any limit to the potential contribution of a clone; can it make any part of the pattern within the compass of its growth? To answer this question the technique has been modified so that the marked clone grows much more than the other, unmarked, cells (Figure B4.1B). This 'Minute technique' led to the discovery of competition (p. 148) and compartments (see The segmentation genes, p. 201).

Clones can be generated in the germ cells of females so that the marked cells can form a clonal group of eggs. For example, females heterozygous for a dominant female sterile can be irradiated as larvae. This can generate a clone of germ cells that are homozygous for the wildtype allele of the dominant sterile and are therefore competent to form normal eggs. Such females, after irradiation, do lay clonal groups of fertile eggs.

All the methods listed above can be modified in an important way; the marker mutant can be linked to a mutation of interest. Thus eggs homozygous for *Notch⁻* can be made by making females that are *Notch⁻/Dominant female sterile* and irradiating them as larvae. The clones of viable eggs produced will have developed in the absence of the *Notch* gene and, when fertilised with $Notch^-$ sperm, will make mutant embryos — the phenotype of these embryos can be compared with normal $Notch^-/Notch^-$ embryos as an assay for a maternal contribution from the $Notch$ gene (see p. 6, Figure 1.2).

Another example is the bithorax complex (see p. 111). Embryos were made as shown in Figure 5.2 and irradiated to find out where the wildtype gene Ubx^+ is required. Note the advantage of this: Ubx^- is a lethal allele, Ubx^-/Ubx^- embryos die when they have made a highly aberrant cuticle and fail to hatch. Nevertheless, one wants to know where the Ubx gene is required in the adult and what happens when the gene is removed from adult cells. This can only be discovered piecemeal by removing the gene from areas bit by bit. When Ubx^+ is removed from any part of the antenna or the first leg they develop perfectly normally without it. When the gene is removed from the T3a compartment the cells lacking it are transformed so they now make pattern appropriate to T2a. Thus one can build up, on a map of the adult, a picture of where Ubx^+ acts and what it does — something that cannot be done without making mosaics.

There is one caveat: if clones are normal in an organ it cannot always be concluded that the wildtype gene is not required in that organ. The mutant clone may be rescued by wildtype cells nearby or elsewhere. Also, there is the possibility that the wildtype allele may have made enough product to perform its function even after the gene itself has been removed by mitotic recombination. If this does happen the function is said to perdure. Perdurance is usually short, but for mapping where genes are required it is wise to irradiate young embryos and therefore give the mutant clone a long time to grow and dilute out wildtype gene product.

Box 4.2 Genetic mosaics — cell markers

Mitotic recombination depends on having good cell markers. They should be easy to score, so that wildtype and mutant cells can be distinguished one by one. They should be autonomous, meaning that all cells carrying the marker genotype should express the phenotype and no others — within each cell, genotype should determine phenotype. Nonautonomy does occur — in mosaics made with some wildtype genes the product spreads and rescues the phenotype of nearby mutant cells: such mutations are not used as cell markers. Good cell markers should be gratuitous, meaning they should mark cells but not affect or damage them in any other way. They should, as we have just seen, have a short perdurance, so that small clones can be studied. Most markers affect the cuticle, such as *multiple wing hairs* (see Figure B4.1, Box 4.1) or *yellow*; there are many that change the eye colour, such as *white* and these may be scoreable in both the pigment cells and the photoreceptor cells of the eye.

The X-ray induced recombination must occur between the centromere and the marker of interest. This is illustrated in a particular case in Figure B4.2, featuring the cell marker *pawn* (*pwn*). When breakage and rejoining occurs at position 1, clones of cells are marked with *pawn* and these will also be *engrailed⁻* (*en⁻*) and *Minute(2)c⁺*. At position 2, unmarked *pawn⁺ engrailed⁻ Minute(2)c⁻* clones will be generated and, at position 3,

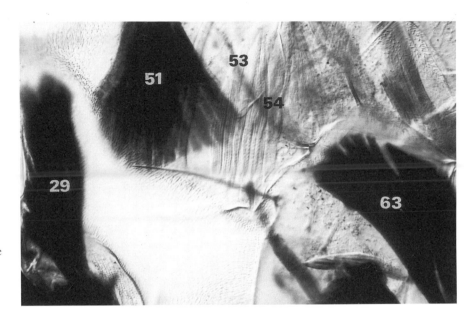

Figure B4.2 Making marked *engrailed⁻* clones that are also *Minute⁺*.

Figure B4.3 A *sdh^{ts}* clone in some wing muscles. The clone has colonised the dorsal muscles which include muscle numbers 53 and 54, but not 51.

Minute(2)c⁺, but otherwise wildtype clones will be made. In the three examples the sister cells from the exchange will be *Minute(2)c⁻ / Minute(2)c⁻* and will die.

In recent years the arrival of transposable elements that carry wildtype genes has proved useful; elements carrying *white⁺* have been moved around the genome and this allows most mutants that affect eye development to be looked at as *white⁻* clones in a *white⁺* background.

Markers for cells that are inside the body — in the central nervous system or other organs — are not so well represented. There are mutants which lack enzymes and these can be used if stained histochemically or with antibodies. A temperature-sensitive form of the succinate dehydrogenase protein is found in flies homozygous for the *sdh^{ts}* mutation. These flies fail to develop at 25°C but breed normally at 20°C. When heterozygotes are irradiated and the flies stained for succinate dehydrogenase, clones of *sdh^{ts}* cells can be easily seen in most internal organs. An example is shown in Figure B4.3 which is a dissection of the wing hinge region showing some muscles (if you wish to see these muscles in their setting, compare with Figure 4.5). The clone is *Minute⁺/Minute⁺* in a *Minute⁺/Minute⁻* background; it grows excessively, filling the dorsal set of T2 muscles which do not stain for succinate dehydrogenase. Outside the clone the genotype of the muscles is *sdh⁺/sdh^{ts}* and they stain dark blue.

For more information, see Roberts (1985) (details p. 22).

The link with genetics

The idea of a compartment is central to thinking about making a fly and this is mainly because it provides a link of understanding between the genes that are responsible for design of organs and the groups of cells that construct those organs. The history of the *engrailed* gene is described on p. 207, but here we can state that *engrailed* is expressed and required in all the cells of all posterior compartments. Like many statements it can be essentially true, even if there are some exceptions — for biology is full of minor variations and subroutines. The statement rests on a number of experiments but the most decisive is that which demonstrates requirement. To test for requirement one removes the wildtype allele of the gene from a cell and its descendent clone. If there is no effect on pattern one concludes that the gene is not needed in the region occupied by the clone. The plan of the experiment for *engrailed* is shown in Figure 4.3. Look at A: in the wildtype wing the compartment boundary is precisely positioned. There is a weak allele called *engrailed¹*, which is caused by a large insert into the regulatory region, 3' to the protein coding region of the gene. Wings homozygous for *engrailed¹* have no precisely defined compartment boundary. In B we see this is due to the effects of the mutation on posterior cells — for although anterior clones of *engrailed¹* cells are completely normal, posterior cells make abnormal patterns and even cross the line where the border should be. Strong *engrailed* alleles that are lethal to the whole animal can nevertheless be studied in cell clones (Figure 4.3C). Such anterior *engrailed^L* clones are again completely normal while posterior clones make defective patterns and cross over into anterior territory. Sometimes the posterior clone becomes almost completely segregated into the anterior compartment and now respects an irregular line near to where the compartment boundary should be, but from the anterior side (arrow). The results on the wing are clear cut; removal of the wildtype *engrailed* allele from all anterior cells has no effect, removal from any posterior cells changes those cells and only those. These changes are partial — the posterior cells acquire only some anterior properties — for example they make sensory structures that are normally found only in the anterior wing but they are not arranged in the normal anterior pattern.

These experiments show that the *engrailed* gene is responsible for the miscibility of posterior cells during growth. These mixing properties, or 'affinities' of cells, are important but little understood. In normal growth of the wing the cells divide and intermingle a little with their neighbours, but do not let go of their sister cells. This means that, over time, a clone remains coherent but the boundary becomes jagged. The cells behave differently across the compartment border where the degree of intermingling during growth is reduced; cells of anterior type and of

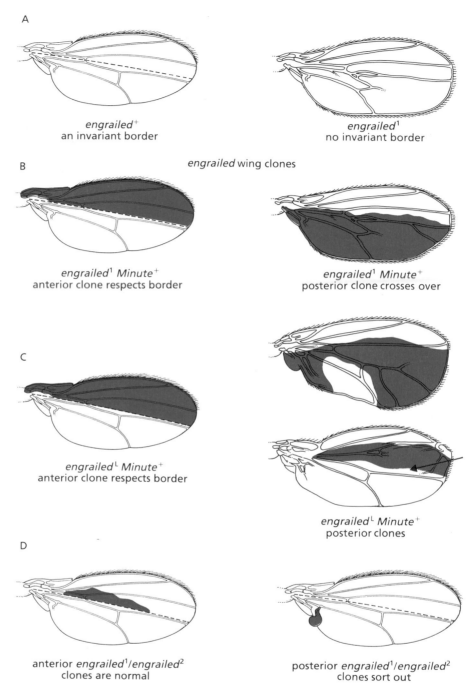

A

engrailed[+]
an invariant border

engrailed[1]
no invariant border

engrailed wing clones

B

engrailed[1] *Minute*[+]
anterior clone respects border

engrailed[1] *Minute*[+]
posterior clone crosses over

C

engrailed[L] *Minute*[+]
anterior clone respects border

engrailed[L] *Minute*[+]
posterior clones

D

anterior *engrailed*[1]/*engrailed*[2]
clones are normal

posterior *engrailed*[1]/*engrailed*[2]
clones sort out

Figure 4.3 The behaviour of wing clones that are defective for the *engrailed* gene.

posterior type hold on more strongly to cells of their own class and, in effect, minimise the contact with cells of the other. The result is a relatively straight line between the two, like the interface between oil and water, and this is maintained even when cells on one side of the compartment boundary divide more rapidly than those on the other.

89 CELL LINEAGE AND ALLOCATION

However, if the posterior cells are mutant for *engrailed*, they now acquire anterior adhesive properties and, if they are near the compartment border, they can cross over into anterior territory, displacing anterior cells as they do so (Figure 4.3B and C). When a posterior clone carrying a strong mutant allele of *engrailed* is made away from the compartment border, it acquires anterior-like affinities. This results in the boundary around the clone being minimised more forcefully and, eventually, a narrowing ring pinches out the clone cleanly so that it separates from the epithelium as a discrete vesicle (Figure 4.3D). All these experiments serve to illustrate that anterior and posterior cells are intrinsically different, a difference depending on the *engrailed* gene.

Marking embryonic cells in legs, the proboscis and elsewhere, suggested that the embryo consists of a chain of alternating anterior and posterior compartments and when the *engrailed* gene was cloned (see *engrailed*: the life history of a gene, p. 207) and antibodies made, this picture in the mind became realised under the microscope. Figure 4.4 shows an embryo at stage 11 stained with a monoclonal antibody against a small part of the *engrailed* protein. The parasegments (see below) are demarcated by surface grooves and are numbered; the *engrailed* protein is found in epidermal nuclei in the anterior portion of each parasegment. Underlying those nuclei one can see that the central nervous system also contains stripes of neuroblasts containing the *engrailed* protein. Estimates from gynandromorphs and clone frequencies had led to the expectation that the posterior compartments

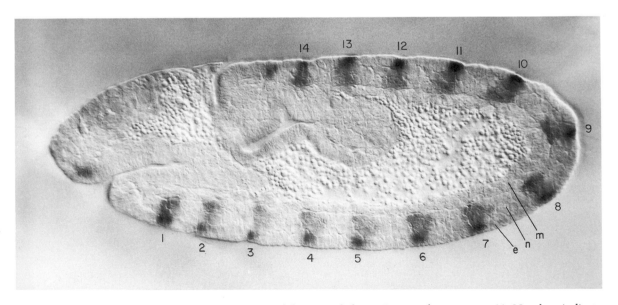

Figure 4.4 Expression of the *engrailed* gene in an embryo at stage 11. Numbers indicate the parasegments; e, epidermis; n, neuroblasts, m, mesoderm.

would be less than half the size of the anterior ones, and this is reflected in the cells labelling with *engrailed* being about one-third to one-quarter of the segment width. All the evidence from cell lineage, *engrailed* requirement and *engrailed* expression agrees very nicely and points to the epidermis of the embryo as being divided up into cells of two types arranged in alternating stripes.

Segments and parasegments

There used to be no certain way of demarcating a segment in arthropods and there were many disputes about the segmental provenance of a particular organ, or the position of a segment boundary. The most reliable criterion was the connections made between muscles and cuticle — many muscles attach near to the segment boundary — but the resolution is low. The evidence that the segment boundary is a compartment boundary and delineates an interface between two different sets of cells came from cell marking and grafting experiments on a bug, *Oncopeltus*. The modern way to define a segment of the ectoderm is as a pair of compartments, one anterior and one posterior.

Traditionally, the segment has been considered the unit of construction, but tradition is not evidence. Indeed it now seems that the parasegment, not the segment, is the **developmental** unit of importance in the embryo. Parasegments are defined early in development, and each is a precise set of cells. They are out of phase with segments; in the presumptive epidermis a segment is one pair of compartments anterior and posterior, a parasegment is the alternate and overlapping pair, posterior and anterior. Parasegments extend into the mesoderm (see Figure 1.6). The main body of evidence that parasegments are first and foremost in insect design comes from the function and expression of selector genes (see Chapter 5), but the argument has other components. For example, the first anatomical signs of metamerisation in the insect embryo are little grooves that appear after the germ band has extended (see Figures 1.4 (p. 12) and 4.4). These grooves are at the boundaries between parasegments. Also, the chain of parasegments seems to contain 14 complete ones — in the central nervous system the chain starts with a posterior compartment and ends with an anterior one. If the chain had been made of 14 segments it would have begun with an anterior and ended with a posterior compartment. The strength of this argument is weakened when, later on in development, it seems as if more compartments are added to either end of the chain.

More persuasive, that is less subjective, is the evidence (pp. 97, 101) that the first allocation of embryonic cells is to parasegments and that these are only **subsequently** divided up with the help of the *engrailed* gene into anterior and posterior compartments. In other words, parasegments and compartments definitely exist as entities in the embryo's 'internal description' while segments, as developmental units of design in the embryo, may only exist in the mind of the scientist (Box 4.3).

Box 4.3 External and internal description

'Inside every animal there is an internal description of that animal', Brenner said. Some of the hairs in his eyebrows were 3 inches long. 'We do not yet know what the *names* are. What does the organism name *to itself*? We cannot say that an organism has, for example, a name for a finger. There's no guarantee that in making a hand, the explanation can be couched in the terms we use for making a glove.

'What is a mouse? ... it may be the only way to give an answer to this is to specify an algorithm for how you could build a mouse. **In the way the mouse builds itself**. And you must be careful not just to give a description of the mouse as it exists ... in biology, programmatic explanations will be algorithmic explanations. You will have to say, Next switch on gene group number fifty-eight. And then one has a whole lot of molecular biology — what is gene group fifty-eight and what does it do.'[7]

There are two images of a fly; one is that of the anatomist who gives names to parts. These parts will be allocated names on various bases including anthropomorphism; for example the heart is named as such because it is an organ defined by structure and function. The eye is more tricky; do we mean only the part of the eye that is on the surface or do we include the lamina of the brain which underlies it? And where, exactly, does the eye end and the head begin?

The other image of a fly is derived from an attempt to see how the fly is constructed; it is an atlas of names based on developmental anatomy rather than the anatomical outcome. Consider the midgut — this is the middle part of the gut and is given one name by the traditional anatomist. However, the midgut develops by the fusion of two primordia that come from two parts of the embryo that could not be more remote, so the anterior and posterior midgut might deserve different names. The reason for attempting to move towards a developmental nomenclature is simply that it is more objective and exact. When developmental criteria are used, organs can be better defined because, usually, they are precisely delimited. This has long been recognised as a worthwhile aim and developmental criteria have been preferred, when available, to others. An example is the ageless dispute about whether the malpighian tubules 'belong' to the hindgut or the midgut. This is now being solved by embryological and genetic evidence which places the malpighian primordial cells in the ectodermal hindgut.

There is a danger of disputes like this seeming old fashioned and sterile and they are, so long as the aim is unclear. If the aim is to understand the 'internal description' of the fly, that is to provide a picture of the way structure and pattern are encoded in the genetic information, the way three dimensions arise out of one, then it is worthwhile.

Segmentation of the mesoderm

Cell lineage studies can be done on the mesoderm, just as on the ectoderm; the ideas and approach are the same, the methods are somewhat different. Succinic dehydrogenase is an enzyme that can be detected with a blue stain. Cells of a mutant called *succinate dehydrogenase*[ts] (*sdh*[ts]) lack this enzyme and, in clones of cells, will show up as white patches on a blue background. As with the ectoderm, the embryonic mesoderm cells divide only once or twice before the larval muscles are formed, so little has been learned from mitotic recombination of larval mesoderm. However, adult muscles do derive from founder cells that undergo many divisions and large clones of *sdh*[ts] muscles can be induced. When *Minute*[+] clones are made by irradiating blastoderm embryos, the clones always remain confined to single segments in the thorax, filling precisely defined sets of muscles. If thoracic clones are initiated by irradiation of young larvae, the muscle precursors are apparently confined to particular imaginal discs or to the neighbourhood of the spiracle. Mitotic recombination clones and nuclear transplantation have been combined to make Figure 4.5 which classifies the adult muscles of the thorax according to their origin. Parasegments 3–5 contribute to these muscles. There are three main sets that correspond to the segments, grey for T1, red for T2 and brown for T3. Although the muscles can be divided into dorsal and ventral sets (the dorsal sets are shown in lighter colour), there is no evidence for anterior and posterior compartments. This brings the mesodermal parasegments into simple registration with the ectodermal segments; for example, I believe the somatic .mesoderm cells of parasegment 4 entirely or largely form the adult muscles that belong to the single T2 set shown in Figure 4.5, as well as a larval set of T2 muscles. Likewise, parasegment 5 forms the T3 set of muscles. Support for this picture comes from the patterns of expression of selector genes (e.g. *Ubx*, p. 121) but there may be exceptions.

In the young embryo the dorsal and ventral subsets of muscles in each segment come from a single pool of cells, but soon these precursors become separated from each other and from the two spiracular muscles. Note that the muscles 80 and 81 link the thorax and abdomen and illustrate that the origin of a muscle cannot be deduced from its attachment sites — for although they attach to both T3 and A1, they originate only from A1. Other examples of this are given by muscle numbers 48 and 66, both of which span between wing and leg cuticle; the former comes from the dorsal wing set of muscles while the latter comes from the ventral leg set.

Isolated mesodermal cells are capable of fusing indiscriminately; if genetically marked (*sdh*[+]) muscle precursors from the T2 wing disc are released into the blood of a *sdh*[ts] host, they will contribute to many muscle fibres, including those in the abdomen. This clearly does not

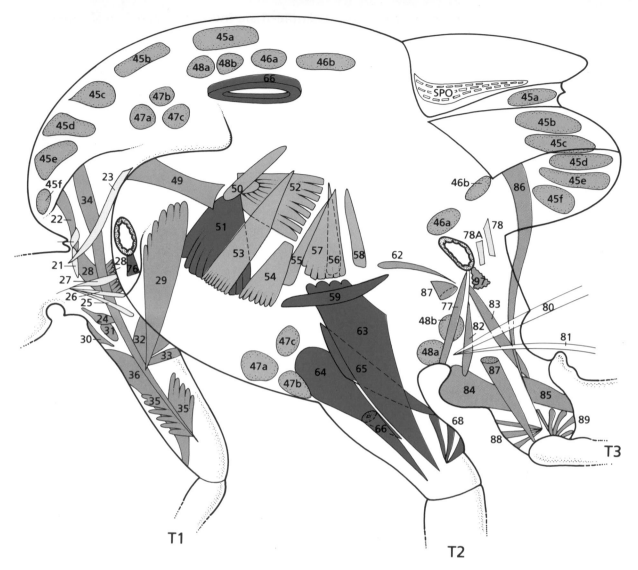

Figure 4.5 The muscles of the adult thorax, coloured according to origin. The scutellar pulsatile organ (SPO) is part of the heart muscular system.

occur to any extent normally or the segregation of muscles into different compartments by lineage would not have been observed. Presumably, even in the embryo (for the irradiation experiments and nuclear transplant experiments mark embryonic cells) the mesodermal cells are tethered to each other in loose sheets and are not free to wander off. Later on, the sets of mesoderm cells become sequestered in discs and near spiracles. How determined are individual mesodermal cells? The experiments described above pull the answer to this question in opposite directions: lineage and selector gene expression would indicate that mesodermal cells in different parasegments are intrinsically different, while the experiments on selectivity of fusion suggest they are not. The answer may be found by studying muscle development: it seems that each muscle is initiated by a specific founder cell and, probably, these become identified by which parasegment they belong to and their position within it. Later, mesodermal cells from a general pool fuse with the growing founder cell and it may be that these cells are naïve, acquiring their genetic state from the cytoplasm they enter. There is circumstantial evidence for this model: to form a specific muscle, single cells expressing the homeodomain protein S59 fuse with a few other mesodermal cells that do not express S59. As the cytoplasms fuse, antibody staining shows that the new nuclei begin to bind the S59 protein which at first diminishes in the nucleus of the founding cell, but soon recovers and then all the nuclei stain strongly. If the S59 protein is indeed playing a role in the identity of the muscle (as would seem very likely) then the nuclei that join the original founder acquire its identity.

How are the cells allocated to parasegments?

Chapters 2 and 3 outlined how a gradient of *bicoid* protein becomes transformed, step by step, into the striped expression of the pair rule genes. One role of these stripes is to allocate cells to parasegments.

I shall consider cell allocation mainly in the anteroposterior axis, because the situation in the dorsoventral axis is less clear. By the blastoderm stage, in both axes, there are many stripes of cells expressing transcription factors. In the case of the pair rule genes the stripes are carefully positioned, but still have graded edges. When you look at Plate 4.1 you see a pattern that changes steadily over time, you see the brown *ftz* stripes and the grey *eve* stripes emerging as strong features out of a more even field of expression. These stripes get stronger and stronger as they narrow and clear interstripes appear. The stripes become asymmetric, developing sharp, intense anterior boundaries but still having hazier, indefinite posterior borders. It is very beautiful, for I know of no better system to watch the process of pattern formation actually happening. (There are other examples in nature, such as the acquisition of colour in a butterfly's wing as the scales mature, but there you see the preexisting pattern developing like a photographic print, you do not see the pattern itself forming.)

As you look at those pictures an important question emerges: when does the pattern matter; are there one or more times when it is utilised to produce the next steps in the formation of the final pattern? The expression of many genes has now been described, and it is strange this question is rarely asked, for many of the patterns seen are ephemeral and may only be steps on the way to an effective pattern, or decaying relics of it. In the case of *ftz* and *eve* the assumption has usually been made that the stripes are effective when they are about equal in width to the interstripes. Under this model, at a critical time all cells expressing *ftz* would be allocated to even-numbered parasegments and those expressing *eve* to odd-numbered parasegments. This is an attractively simple hypothesis but unfortunately for it, the stripes are never exactly reciprocal — they still overlap and have fuzzy boundaries at a time when gaps appear between them (Plate 4.1). Yet the data on cell lineage (p. 79) make it clear that parasegments must be exactly defined with no cells left unallocated. So how can one try to determine when a changing pattern of gene expression counts? In the case of *ftz* and *eve* it is my guess that the right time is likely to be when products of these genes reach their highest concentrations in some of the cells. At the same time, the anterior margins become stabilised and sharp and these are tantamount to the parasegment borders (see below). Both these events occur at, or near, the beginning of gastrulation, at stage 6 (Figure 1.4, p. 12) and, probably, it is from this time that cells are progressively allocated first to parasegments and then to anterior and posterior compartments.

The regulation of the *eve* gene can be divided into two phases, each depending on a different part of the *eve* upstream region. Each *eve* stripe is first placed in the embryo rather independently by regulatory regions which respond to concentration gradients of gap gene proteins. The subsequent sharpening of the *eve* stripes is a special process that depends on a particular 5′ piece of DNA, a 'late' element placed within about 800 base pairs upstream of the gene. This is shown by placing that element upstream of the reporter gene *lacZ* and producing sharp stripes of β-galactosidase with definite and stable anterior margins. The formation of these sharp margins depends on other genes including the *hairy* gene and even *eve* itself. In *hairy*⁻ embryos the later sharpening does not occur. In *eve*⁻ embryos the stripes do not sharpen properly, so it follows that there is an autoregulatory feedback loop: one might expect *eve* binding sites to be found in the late element. It is not clear what makes the anterior cells especially liable to become locked into this autoregulatory loop, but it may be *hairy* since it is coexpressed in the anterior part of the *eve* stripes.

It is clear that the anterior margins of *eve* and *ftz* stripes are an important and stable feature of the stripes because they can be mapped relative to *engrailed* expression. The mapping is not straightforward, because it is necessary to have both *engrailed* and the two pair rule genes strongly expressed at the same time and, normally, the *ftz* and *eve* stripes are fading while the *engrailed* stripes are coming up. Also, the *ftz* protein only has a half life of about 5 minutes, so it is difficult to be certain that there is a stable anterior margin and, if so, exactly where it is. Both these problems are solved by using hybrid genes which are transformed into flies in which sufficient 5' region of the *ftz* or *eve* DNA drives expression of *lacZ*; the *lacZ* gene is expressed in exactly those cells which transcribe the *ftz* or *eve* coding regions in the same fly. The advantage of the β-galactosidase is that it persists for a few hours; it shows that there is a sharp and stable anterior margin and marks the cells delineated by it. These anterior margins are then compared cell by cell with the expression of *engrailed*. When you do this it becomes clear that the anterior margins of the *ftz* and *eve* stripes coincide exactly with the anterior margins of the *engrailed* stripes and are, therefore, the parasegment borders (Figure 4.6). This cell-by-cell coincidence can best be explained if the anterior boundaries of the *ftz* and *eve* stripes are actually **responsible** for activating the *engrailed* gene in the border cells. These same boundaries therefore delimit the parasegments by allocating cells to one or the other. Each parasegment consists of just those cells that lie between the anterior border of the one stripe and the anterior border of the next. Thus, the cells of parasegment 6 can be identified in the embryo as those that lie between the sharp anterior edge of the third *ftz* stripe and the sharp anterior edge of the fourth *eve* stripe (Plate 4.1 and Figure 4.6). Within those, the *engrailed* gene subdivides the cells of the parasegment into posterior and anterior subsets. The pattern of *engrailed* expression takes time to develop and is not definitive until some time after the parasegment boundaries are drawn. Only then are the segment boundaries delineated and the compartments founded.

How is it known that the *ftz* gene directs the *engrailed* gene rather than vice versa? The answer depends on molecular epistasis (see Box 2.4, p. 45) — in *ftz*⁻ embryos, *engrailed* protein is absent in the even-numbered parasegments, while in *engrailed*⁻ embryos, *ftz* expression is normal (Figure 4.7).

While, topologically, this model makes sense, it is not clear what the mechanics of it are. How does *engrailed* come on with respect to *ftz* and why is it only in the anterior one-third to one-quarter of the parasegment? The answer to this question is not known — but it is important to know. Important because, although both *ftz* and *engrailed* are homeobox-containing transcription factors, *ftz* will soon disappear while *engrailed* will remain on in the posterior cells and their descendants right through to the adult. This means that *engrailed* is the first gene known whose distribution amounts to a permanent chart of the body plan of the animal — at the cellular level. *engrailed* begins to come on at the end of the blastoderm period, in some parasegments before others, and is active in all its 14 stripes in the gastrulating embryo. Since the border of the parasegment as defined by *ftz* and *eve* is wiggly,

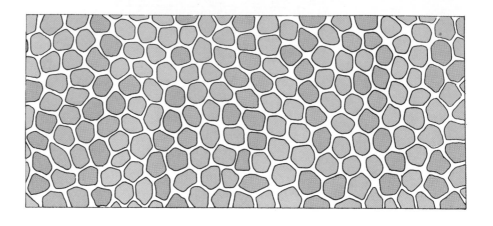

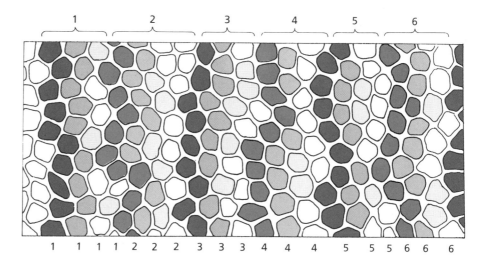

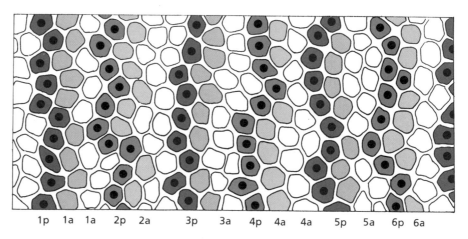

Figure 4.6 The maturation of the *ftz* (red) and *eve* (grey) stripes and the allocation of compartments. The parasegment boundaries are delineated by the anterior boundaries of the stripes; later the anteriormost cells begin to express *engrailed* (central black dot).

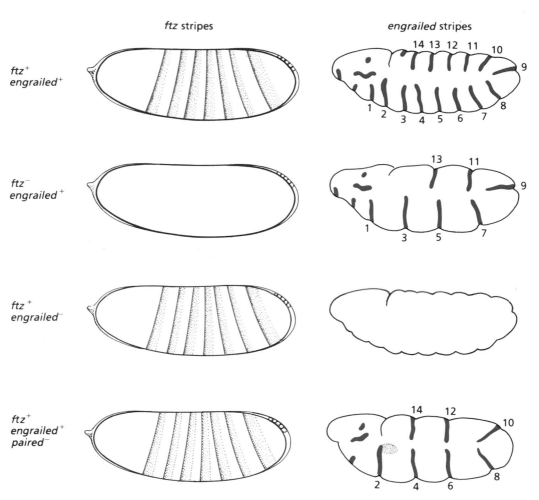

Figure 4.7 Molecular epistasis: *engrailed* expression depends on *ftz* expression; *ftz* expression is independent of *engrailed* and *paired*.

and since the expression of *engrailed* follows that border so exactly, it is likely *engrailed* depends on the *ftz* and *eve* proteins directly. Indeed, there are *ftz* and *eve* binding sites in the *engrailed* promoter and experiments *in vitro* support this idea. This hypothesis predicts that if *ftz* were expressed in the wrong cells, and at a high level, *engrailed* could also be activated. And this is so: some cells in embryos carrying the *ftz* gene under heat shock control do get locked into *ftz* expression and these cells do switch on *engrailed*.

Why do only the anterior subsets of the cells expressing *ftz* and *eve* come to express *engrailed*? One theory might be that there is more *ftz* or *eve* near the anterior boundary, there is a gradient within the stripe, and a threshold concentration exists, above which *engrailed* would be activated. If this were the case, increasing the total amount of *ftz* product should broaden the *engrailed* stripe and experiments suggest this does not occur. Another possibility is a combinatorial one; maybe there is a different gene whose expression pattern overlaps only partly with the *ftz* or *eve* stripe and this gene has a sharp posterior border. Cells would turn on *engrailed* if they lay between the sharp anterior border of *ftz* or *eve* and the sharp posterior border of this gene. A possible candidate (for the *eve* stripes) is the homeobox gene *paired*: *paired⁻* embryos lack the odd-numbered *engrailed* stripes (Figure 4.7) and the *paired* pattern of expression is consistent with this; the stripes of *paired* protein overlap only partially with the *eve* stripes, extending about one or two cells posterior to the anterior margins of the *eve* stripes. It will be necessary to look at the pattern of *engrailed* expression *vis-à-vis* both *eve* and *paired* to see if the model holds up at cellular resolution.

There are several combinatorial models in the literature; a recent one points to the possible role of repression — perhaps a gene might be expressed in only the posterior part of the *ftz* stripes and prohibit *engrailed* expression there. A candidate gene has been identified; it is called *odd-skipped*, encodes a zinc finger protein and is expressed in about the right region, overlapping the posterior regions of the *ftz* stripes. It has not yet been studied at cellular resolution.

Still another hypothesis takes us more into the unknown; perhaps gradient fields of positional information (see Chapter 6) are demarcated by the nascent parasegment boundaries. Here, high levels of the *ftz* and *eve* proteins might fix the anterior boundaries of *engrailed* expression while the posterior boundaries of the *engrailed* stripes could be determined later and depend on a threshold response at a particular gradient value. Under this hypothesis the parasegment borders would exist before the segment boundaries. There are a number of problems with this rather woolly perspective, but they will have to be faced: at some unknown point in development the gradients of positional information are established and must eventually take over the patterning of the segment. Some support for this hypothesis comes from locust embryos, where *engrailed* is activated **progressively**, first in cells at the parasegment borders and then backwards from them. Such an observation seems more in key with a gradient model than a digital and combinatorial one — but the argument is weak. It is going to be important to find out exactly how *engrailed* is switched on in the right cells.

I have not discussed the other pair rule genes whose function is required for the process of cell allocation to parasegments (an example is *runt*). The reason is that, apart from their positions in the hierarchy, there is not enough known about them. Some have been cloned and some not and for the moment they do not easily fit into the story. It is not as if the reader is short of genes to think about! There will be others that did not appear in the screen of Nüsslein-Volhard and Wieschaus. These may play key roles in the process, but their products might not need to be localised to particular cells. In these cases it may well be that the gene product is deposited by the mother in the egg and is sufficient to sponsor development past the process of segmentation. Only removal of these genes from the germ cells can reveal that they do function in segmentation.

Segment polarity genes

Many of the pair rule genes are transcription factors and control the expression of other genes, including the class of 'segment polarity genes'. Segment polarity genes are defined by their mutant phenotype — they are lethals which change the pattern, and often the polarity, of **every** segment. There is accumulating evidence that their wildtype role is to pattern the segment, or at least to provide the necessary cellular machinery — such as intercellular junctions — so that patterning can occur.

When segment polarity genes come into play the parasegment borders are already delineated by *ftz* and *eve* and the *engrailed* gene is becoming activated. The segment polarity genes are needed from then on: the parasegment border has to become established as a long-lived feature with specific cellular properties. The activation of the *engrailed* gene has to mature from a provisional pattern into a definitive one: eventually, after several steps, those cells expressing *engrailed* become locked into an autoregulatory feedback loop.

In the wildtype ventral abdomen each segment is divided into an anterior region of denticles and a posterior region which is naked. The posterior compartment lies almost entirely in the naked region but includes the most anterior row of denticles (see Figure 1.6, p. 19). The denticles are oriented (Figure 4.8), showing that the cells are polarised (see Chapter 6). Mutations in the segment polarity genes transform this pattern; the biggest class of mutant phenotypes is epitomised by *wingless⁻*. In *wingless⁻* embryos, the whole of the ventral abdomen is covered by denticles which still show weak polarity — the normal polarity is found in the anterior half and a reversed one in the posterior (Figure 4.8). It is as if the whole segment now differentiates at one local level in the normal pattern. There is no naked cuticle, there is no posterior compartment and, when the pattern is mature, there is no expression of *engrailed*. This pattern would seem to be a kind of default state, for a number of genes give a similar mutant phenotype: examples are *armadillo*, *gooseberry* and *hedgehog*. There is a growing list of genes which also give the same phenotype when mutant,

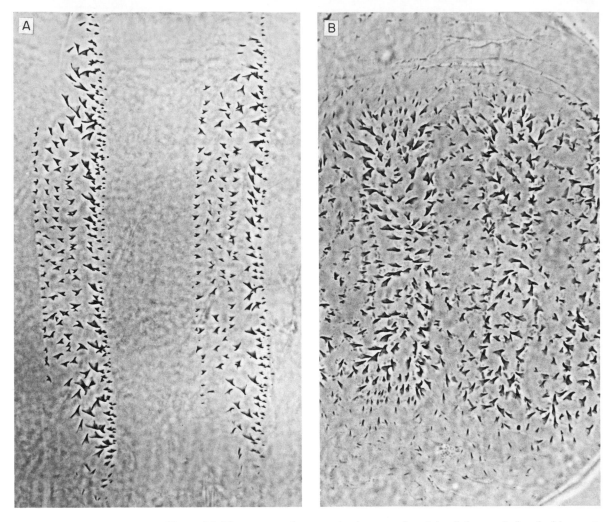

Figure 4.8 The segmental pattern in the ventral cuticle of the newly hatched larva. A, two denticle bands of the wildtype abdominal segments (A4 and A5). B, equivalent region of a *wingless⁻* larva. Compare with Figure 1.6, p. 19.

but only when the maternal contribution is removed (*dishevelled*, *porcupine*, *fused*).

If the *wingless* gene is expressed everywhere, under heat shock control, the result is a naked embryo with no denticles at all. The new pattern does not correspond to one level in the segment, because half the segment expresses *engrailed* and half does not. One particular segment polarity mutation gives exactly this phenotype, the gene is appropriately called *naked*.

Two other mutations give an apparently unrelated phenotype: there is a double band of denticles, the posterior half being oriented in reverse. There are two stripes of *engrailed* and correspondingly two segment boundaries. These genes are called *patched* and *costal-2*.

Understanding segment polarity genes is proving difficult. Some of these genes have been cloned and one, *armadillo*, encodes a protein that in vertebrates is required to make cell junctions. *wingless* makes a cysteine-rich protein that is secreted and is taken up by nearby cells, it is homologous to the vertebrate oncogene, *int-1*. The protein seems to become associated with extracellular material but it is not known whether it is processed and whether there is a receptor for it. Initially *wingless* is expressed in a row of cells exactly adjacent and anterior to the *engrailed*-expressing cells; both *engrailed* and *wingless* therefore define the parasegment border and both are needed, directly or indirectly, for the expression of the other. In *wingless*⁻ embryos, *engrailed* expression begins normally (in accord with the idea that it is fixed by the pair rule genes) but fades away. Likewise in *engrailed*⁻ embryos, *wingless* expression is initiated as usual but decays. In both *patched*⁻ and *naked*⁻ the *wingless* and *engrailed* patterns are altered; changes do not affect the parasegment border, for example in *naked*⁻, the *engrailed* stripe spreads out from that border.

Broadly, there are two extreme ways of depicting the wildtype functions of these genes. First, they could be there to label the cells of the developing segment, particularly the anterior compartment. A number of these models have the advantage of being definite; according to one the embryonic segment is divided into just four **discontinuous** cell states. The disadvantage of these models is that they have been 'carpentered' to fit the available data and have little predictive power. Second, is the view that these genes are necessary to establish the segmental gradient of positional information — a concept that owes its existence to experiments on the cuticular polarity and pattern of other insects. The segmental pattern, within each compartment, is seen as the outcome of a **continuum** of cell states that elicits a common polarity (Chapter 6). The disadvantage of this model is that it is uncomfortably vague. It suggests that a secreted molecule such as *wingless* might be a graded morphogen, and other molecules receptors, kinases, proteases, etc. Those other segment polarity genes that have been cloned so far (*patched*, a transmembrane protein; *fused*, a serine/threonine kinase) are not inconsistent with this perspective.

Pattern formation in the early embryo takes advantage of the free diffusion of proteins between nuclei, but after gastrulation this cannot happen because cell membranes have been completed. If segment polarity genes generate mechanisms to channel information across cell membranes they should be useful more generally. Indeed, several of them have mutant phenotypes that cause malformations in the adult and several show intriguing patterns of expression in the imaginal discs. Some have closely related homologues in vertebrates. All reasons why segment polarity genes deserve further study.

Further reading

REVIEWS

Akam, M. (1987) The molecular basis for metameric pattern in the *Drosophila* embryo. *Development* **101**: 1–22.

Crick, F.H.C. and Lawrence, P.A. (1975) Compartments and polyclones in insect development. *Science* **189**: 340–347.

DiNardo, S. and Heemskerk, J. (1990) Molecular and cellular interactions responsible for intrasegmental patterning during *Drosophila* embryogenesis. *Semin. Cell Biol.* **1**: 173–183.

Garcia-Bellido, A., Lawrence, P.A. and Morata, G. (1979) Compartments in animal development. *Sci. Am.* **241**: 102–110.

Ingham, P.W. (1988) *The* molecular genetics of embryonic pattern formation in *Drosophila*. *Nature* **335**: 25–34.

Lawrence, P.A. (1981) The cellular basis of segmentation in insects. *Cell* **26**: 3–10.

Lawrence, P.A. (1988) The present status of the parasegment. *Development* (Supplement) **104**: 61–65.

Morata, G. and Lawrence, P.A. (1977) Homeotic genes, compartments and cell determination in *Drosophila*. *Nature* **265**: 211–216.

Wood, W.B. (ed.) (1988) *The Nematode Caenorhabditis elegans*. Cold Spring Harbor Laboratory Press, New York.

SELECTED PAPERS

Cell lineage

Bryant, P.J. (1970) Cell lineage relationships in the imaginal wing disc of *Drosophila melanogaster*. *Dev. Biol.* **22**: 389–411.

engrailed

DiNardo, S., Kuner, J.M., Theis, J. and O'Farrell, P.H. (1985) Development of embryonic pattern in *D. melanogaster* as revealed by accumulation of the nuclear *engrailed* protein. *Cell* **43**: 59–69.

Hama, C., Ali, Z. and Kornberg, T.B. (1990) Region-specific recombination and expression are directed by portions of the *Drosophila engrailed* promoter. *Genes Dev.* **4**: 1079–1093.

Heemskerk, J., DiNardo, S., Kostriken, R. and O'Farrell, P.H. (1991) Multiple modes of *engrailed* regulation in the progression towards cell fate determination. *Nature* **352**: 404–410.

Lawrence, P.A. and Johnston, P. (1984) On the role of the *engrailed*[+] gene in the internal organs of *Drosophila*. *EMBO J.* **3**: 2839–2844.

Lawrence, P.A. and Struhl, G. (1982) Further studies of the *engrailed* phenotype in *Drosophila*. *EMBO J.* **1**: 827–833.

Morata, G. and Lawrence, P.A. (1975) Control of compartment development by the *engrailed* gene in *Drosophila*. *Nature* **255**: 614–617.

Morata, G., Kornberg, T. and Lawrence, P.A. (1983) The phenotype of *engrailed* mutations in the antenna of *Drosophila*. *Dev. Biol.* **99**: 27–33.

ftz and eve stripes

Carroll, S.B. and Scott, M.P. (1985) Localization of the *fushi tarazu* protein during *Drosophila* embryogenesis. *Cell* **43**: 47–57.
Frasch, M. and Levine, M. (1987) Complementary patterns of *even-skipped* and *fushi tarazu* expression involve their differential regulation by a common set of segmentation genes in *Drosophila*. *Genes Dev.* **1**: 981–995.
Howard, K. and Ingham, P. (1986) Regulatory interactions between the segmentation genes *fushi tarazu*, *hairy*, and *engrailed* in the *Drosophila* blastoderm. *Cell* **44**: 949–957.
Ish-Horowicz, D., Pinchin, S.M., Ingham, P.W. and Gyurkovics, H.G. (1989) Autocatalytic *ftz* activation and metameric instability induced by ectopic *ftz* expression. *Cell* **57**: 223–232.
Jiang, J., Hoey, T. and Levine, M. (1991) Autoregulation of a segmentation gene in *Drosophila*: combinatorial interaction of the *even-skipped* homeo box protein with a distal enhancer element. *Genes Dev.* **5**: 265–277.
Lawrence, P.A. and Johnston, P. (1989) Pattern formation in the *Drosophila* embryo: allocation of cells to parasegments by *even-skipped* and *fushi tarazu*. *Development* **105**: 761–768.

Muscle development

Bate, M. (1990) The embryonic development of larval muscles in *Drosophila*. *Development* **110**: 791–804.
Broadie, K.S. and Bate, M. (1991) The development of adult muscles in *Drosophila*: ablation of identified muscle precursor cells. *Development* **113**: 103–118.
Lawrence, P.A. (1982) Cell lineage of the thoracic muscles of *Drosophila*. *Cell* **29**: 493–503.
Lawrence, P.A. and Johnston, P. (1986) Observations on cell lineage of internal organs of *Drosophila*. *J. Embryol. Exp. Morph.* **91**: 251–266.

odd-skipped

Coulter, D.E., Swaykus, E.A., Beran-Koehn, M.A., Goldberg, D., Wieschaus, E. and Schedl, P. (1990) Molecular analysis of *odd-skipped*, a zinc finger encoding segmentation gene with a novel pair-rule expression pattern. *EMBO J.* **9**: 3795–3804.

S59

Dohrmann, C., Azpiazu, N. and Frasch, M. (1990) A new *Drosophila* homeo box gene is expressed in mesodermal precursor cells of distinct muscles during embryogenesis. *Genes Dev.* **4**: 2098–2111.

SOURCES OF FIGURES

For details, see above.
Figure 4.1 See Crick and Lawrence (1975) and Lawrence (1981).
Figure 4.2 After Bryant (1970). See Crick and Lawrence (1975). Photograph courtesy of G. Morata.
Figure 4.3 See Garcia-Bellido *et al.* (1979), Lawrence and Struhl (1982), Morata and Lawrence (1975) and Morata *et al.* (1983).
Figure 4.4 See DiNardo *et al.* (1985) and Hama *et al.* (1990).

Figure 4.5 After Lawrence (1982).

Figure 4.6 See Frasch and Levine (1987) and Lawrence and Johnston (1989).

Figure 4.7 See Carroll and Scott (1985) and Howard and Ingham (1986).

Figure 4.8 See DiNardo and Heemskerk (1990).

Figure B4.1 After Garcia-Bellido *et al.* (1979).

Figure B4.3 From Lawrence (1982).

Plate 4.1 See Frasch and Levine (1987) and Lawrence and Johnstone (1989) and compare with Figure 4.6.

5 Selector genes

A S THE CELLS ARE ALLOCATED to parasegments, they are also determined as to which parts of the pattern they will produce — antennae or legs, thorax or abdomen. This is done by selector genes, such as those of the bithorax complex, which also define cell affinities. The pattern of muscles depends on selector genes as well as interactions with the epidermis and nervous system.

Diversification of pattern: the selector genes

In normal development many cell types are stable. Once the posterior compartment of the wing has been founded as a small group of cells in the embryo, those cells keep their state of determination and give rise only to posterior wing cells in the adult. Even if the wing imaginal disc is taken from the larva and allowed to grow and grow in culture for years, the cells remember their state — with only very rare lapses (called 'transdetermination'). Such a stable property is important in a system in which a few cells are allocated to specific states (compartments) early on and these must give rise without error to blocks of pattern. It is achieved by a two-step process. First, there is a special class of 'selector' genes which must be activated in the right cells. Second, these genes must be kept locked on throughout the subsequent cell divisions, a process which itself depends on a complex genetic system (e.g. *esc*, *Polycomb*). Selector genes have a controlling role in the sense that they direct the development of the compartment so that the piece of the body pattern constructed is different from that made by another compartment: even though they form homologous modules, the cells allocated to T2a and T3a make different patterns in the cuticle of the larva and adult. It is therefore important that these genes should be switched on where they are wanted and, equally important, switched off where they are not wanted. This concept of selector genes arose out of the discovery of the compartments in the wing and the phenotype of certain mutations in the bithorax complex (BX-C) in which the regions affected coincide exactly with the compartments (see The history of the bithorax complex, p. 211).

    Figure 5.1 illustrates this; it shows the phenotype of three mutations in the regulatory region of the *Ultrabithorax* (*Ubx*) gene, mutations which cause the misexpression of the *Ubx* protein. The *Ubx* gene is

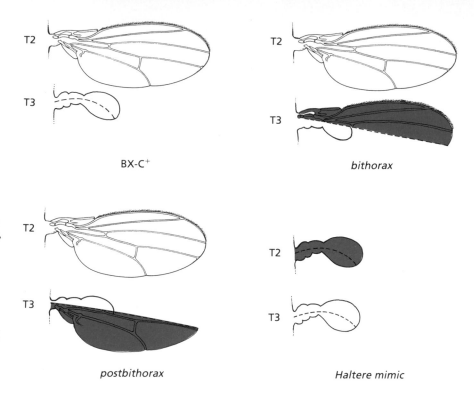

Figure 5.1 Regulatory mutations in the *Ubx* gene. In the wildtype (BX-C+), T2 produces a wing and T3 a balancing organ or haltere. In *bithorax* mutations the anterior compartment of T3 is transformed into an anterior wing compartment. The *postbithorax* mutation transforms posterior T3 to wing while *Haltere mimic*, a gain-of-function mutation, transforms T2 into an extra haltere. Transformed regions are shown in red. When *bithorax* and *postbithorax* mutations are combined the result is a four-winged fly in which T3 is transformed completely into an extra T2.

one of the three genes of the bithorax complex (p. 111). In the wildtype embryo, the *Ubx* gene is expressed in parasegments 5 and 6 and these give rise to T2p, T3a and T3p of the thorax and A1a of the abdomen. Two mutations (*bithorax* and *postbithorax*) transform anterior and posterior halves of the haltere into corresponding regions of the wing (see Plate 5.1). Crucially, the areas transformed are **coextensive** with the compartments as defined by cell lineage. Another mutation, called *Haltere mimic*, causes ectopic expression of the *Ubx* protein in T2 and at a high level, transforming the wing into an extra haltere.

When *Ubx¹* mutant cells are made in the haltere by mitotic recombination, and the cells marked appropriately (Figure 5.2), it is found that every cell in the haltere that loses *Ubx⁺* becomes transformed to wing, showing that the requirement for the *Ubx* gene in the haltere is completely cell autonomous. The part of the wing formed by the *Ubx¹* clone is appropriate for the position of the clone — thus if the clone includes an anterodorsal part of the haltere up to the dorsoventral compartment boundary, then the piece of wing makes the triple row bristles that belong to that part of the dorsal compartment (Figure 5.2). It is as if there were a common ground plan of pattern, such as a system of coordinates, that exists in both the wing and the haltere, in which position is read by the cells in the same way in both organs, but the interpretation depends on the presence of a gene (*Ubx*) which selects differentiation as haltere rather than wing. Although the

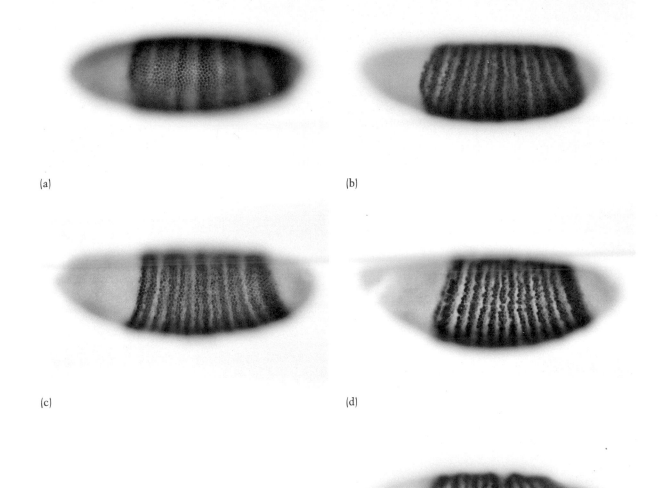

(a)

(b)

(c)

(d)

Plate 4.1 These photographs show the expression of *ftz* (brown) and *eve* (grey) in the cells of the embryo from stages 5(2) to 8(1) — stages are described in Figure 1.4. (a) stage 5(1)–5(2); (b) stage 5(2); (c) stage 5(3); (d) stage 6; (e) stage 8(1). The stripes of *ftz* and *eve* expression narrow from the posterior margin and sharpen anteriorly as they intensify.

(e)

Plate 5.1 Four-winged fly, produced by combining *bithorax* and *postbithorax* mutations (see Figure 5.1).

Plate 7.1 Two butterflies; although they look so dissimilar they are two females of *Papilio phorcas* that differ only in one gene.

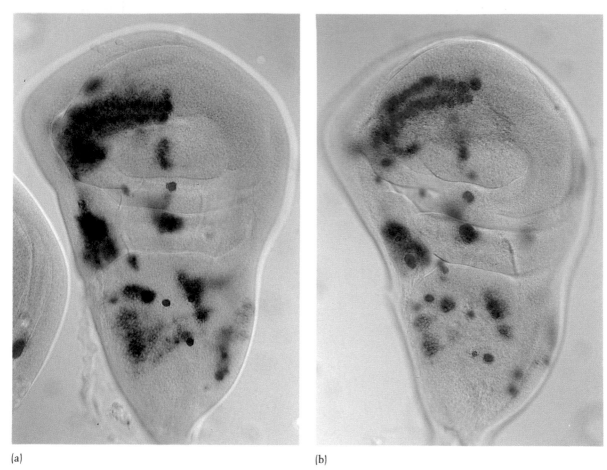

(a)

(b)

Plate 7.2 Expression of the *scute* gene (dark blue stain) in two wing discs. The light blue colour identifies the bristle mother cell.

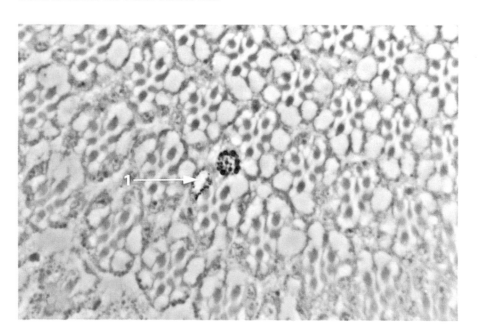

Plate 8.1 An example of a *white*⁺ clone shown in a section of the eye.

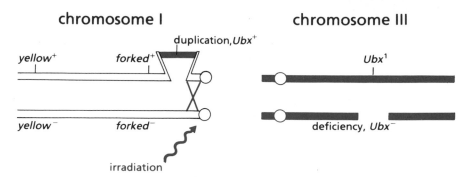

chromosome I

yellow⁺   forked⁺   duplication, *Ubx⁺*

yellow⁻   forked⁻

irradiation

chromosome III

Ubx¹

deficiency, *Ubx⁻*

parent cells are *Ubx⁺*, because of duplication in chromosome I
mitotic recombination in chromosome I leads to loss of *yellow⁺*, *forked⁺* and *Ubx⁺*
giving *Ubx¹*/*Ubx⁻* cells, marked with *yellow⁻* and *forked⁻*

if clone is initiated in anterior haltere (T3a)

Figure 5.2 The *Ubx* gene acts cell-autonomously. The chromosomes of irradiated larvae are shown at top, two examples of clones below. *Ubx⁻* clones transform the cells in the haltere so that they make wing structures.

wing cuticle may form *in situ*

wing cuticle may sort out and form separate vesicle inside haltere

transformation is not complete, a similar picture emerges from the *engrailed¹* clones — here the pattern suggests an underlying mirror image ground plan; clones made on the posterior wing margin make bristles normally found on the anterior margin (see Figure 4.3, p. 89).

In another respect, the *Ubx¹* clones are reminiscent of the *engrailed⁻* clones: they tend to sort out from the surrounding haltere cells. Sometimes the clones are completely pinched out into a vesicle, so that every one of the *Ubx⁻* 'wing' cells is included in the vesicle. Again this shows that selector genes are responsible, not only for the pattern differentiated by the cells, but also for their mixing properties, their affinities.

engrailed is an almost perfect example of a selector gene, for the gene is expressed and required in all posterior compartments. As we have seen, removal of the gene from the whole compartment or from parts of it results in a transformation towards the pattern of another compartment. But the transformation is not complete, as would be expected from the simple model that the only determining difference between anterior and posterior compartments is the *engrailed* gene. There are at least two plausible explanations: there may be other selector genes, in addition to *engrailed*, that are responsible for the difference between anterior and posterior compartments or, less likely, perhaps the removal of *engrailed* gene function is not complete because the mutations used are not nulls.

The idealised selector gene has the following properties: it is active in a compartment or a specific set of compartments and is responsible in a cell autonomous fashion for the pattern formed. It is also responsible for the specific cell affinities that ensure that cells from different compartments do not mix. The selector gene is turned on in the founding polyclone when it is designated and remains activated and required until the end of development in all the cells of the compartments. It is permanently inactivated in cells where it must be switched off, so that the combination of active and inactive selector genes in a cell fixes its state of determination. Selector genes have been identified that qualify to varying extents when compared to this ideal, and many of these contain a homeobox.

The homeobox

For history of the homeobox see p. 216. The homeobox was the first of the DNA binding motifs to be identified in *Drosophila* and has proved extremely useful, particularly in defining new genes in *Drosophila* and, much further afield, in many other animals including mammals. The homeobox in the DNA encodes a homeodomain in the protein. The homeodomain is some 60 amino acids; there are probably over 100 different homeodomain proteins in the fly and the genes have been classified into families according to amino acid sequences. There are four amino acids that are conserved in all the sequences so far analysed from animals and overall the conservation is impressive (one homeobox in the frog *Xenopus* has its homologue in *Drosophila* and of the 60 amino acids, 55 are identical). The sequence encodes four main α-helices linked by turns. The third of these helices fits into the major groove of the DNA, the 'recognition helix', and several amino acids on one face of this helix contact the bases of the DNA. The structure has been studied and this has helped identify those amino acids that make particular contacts; not all of these lie inside the third helix. When sequences from different genes are compared, these contacting amino acids are rather variable and may well determine which binding sites the protein seeks out. If one of these amino acids is altered, so are the *in vitro* binding properties of the domain.

Antibodies against homeodomain proteins show that the proteins are localised in the nucleus and *in vitro* studies with the homeodomain from *engrailed* protein (for example) show that it is capable of sequence-specific binding to DNA. Homeodomain proteins, as we have seen with *bicoid*, make footprints on the DNA; and, in suitable systems, act as transcription factors.

The bithorax complex

The bithorax complex (BX-C) is at once the most perplexing and illuminating genetic system yet discovered in *Drosophila*. The picture of the BX-C has changed over the years (see The history of the bithorax complex, p. 211) but I try here to present an up-to-date but **simplified** story.

The BX-C is concerned with directing the pattern of a particular region of the body, extending in the embryo from parasegment 5 to 14. It consists of three homeobox genes, each producing a number of protein products as a result of variable patterns of RNA splicing. The three genes are called *Ultrabithorax* (*Ubx*), *abdominal-A* (*abd-A*) and *Abdominal-B* (*Abd-B*). The genes are expressed in the epidermis, central nervous system, somatic mesoderm and visceral mesoderm (not in the endoderm), but are best understood in the epidermis, so we will start there. Look at the left column in Figure 5.3A. First, consider the larva that lacks all three genes: now parasegments 5−13 all develop the same way, each parasegment differentiating into the pattern of parasegment 4. This means that there is a chain of reiterated compartments in the epidermis: T1p/T2a, T1p/T2a repeated 10 times. Now, add the *Ubx* gene: the result is a larva that contains one parasegment 4, one 5 (T2p/T3a) and eight of parasegment 6. Add the *abd-A* gene: the result is a larva which has the pattern 4, 5, 6, 7, 8 plus five of parasegment 9 (see Figure 5.3). Three important points emerge from this; the first that the *Ubx* gene and the *abd-A* gene work in particular domains − *Ubx* from parasegment 5 back and *abd-A* from parasegment 7 back. The second, that when the *Ubx* gene functions on its own, it changes part of a chain of parasegment 4s into a chain of parasegment 6s, showing that the genes of the BX-C govern parasegmental (not segmental) states. The third is that there is a combinatorial effect: the *abd-A* gene product adds to the *Ubx* function and both contribute to make the parasegment 7, 8 and 9 patterns.

These three principles also apply to the *Abd-B* gene. Its domain is in the posterior part of the embryo from parasegment 10 backwards and it modifies the combined effect of the *Ubx* and *abd-A* genes. The effects of *Abd-B* are varied, being very weak in parasegment 10 and strongest in parasegment 14, where it predominates. In each parasegment the contribution of each of the three genes is different; for example, slight variations in the role of the *Ubx* and *abd-A* genes are responsible for the slight differences between parasegments 7, 8 and 9.

If genes are removed one can see the same principles. For example, in the absence of the *Ubx* gene (Ubx^- $abd-A^+$ $Abd-B^+$) (Figure 5.3), the two parasegments 5 and 6 become converted to parasegment 4. However, parasegments 7−14 now lack the *Ubx* gene product and this has effects on the pattern of the cuticle − for example, some cuticular

A

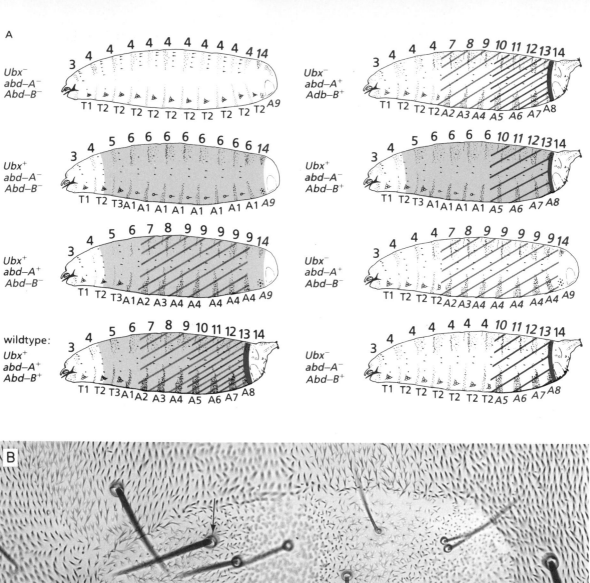

Ubx⁻
abd–A⁻
Abd–B⁻

3   4   4   4   4   4   4   4   4   4   14
T1 T2 T2 T2 T2 T2 T2 T2 T2 T2 A9

Ubx⁺
abd–A⁻
Abd–B⁻

3   4   5   6   6   6   6   6   6   6   14
T1 T2 T3 A1 A1 A1 A1 A1 A1 A1 A9

Ubx⁺
abd–A⁺
Abd–B⁻

3   4   5   6   7   8   9   9   9   9   14
T1 T2 T3 A1 A2 A3 A4 A4 A4 A4 A4 A9

wildtype:
Ubx⁺
abd–A⁺
Abd–B⁺

3   4   5   6   7   8   9 10 11 12 13 14
T1 T2 T3 A1 A2 A3 A4 A5 A6 A7 A8

Ubx⁻
abd–A⁺
Adb–B⁺

3   4   4   7   8   9 10 11 12 13 14
T1 T2 T2 T2 A2 A3 A4 A5 A6 A7 A8

Ubx⁺
abd–A⁻
Abd–B⁺

3   4   5   6   6   6 10 11 12 13 14
T1 T2 T3 A1 A1 A1 A1 A5 A6 A7 A8

Ubx⁻
abd–A⁺
Abd–B⁻

3   4   4   7   8   9   9   9   9   14
T1 T2 T2 T2 A2 A3 A4 A4 A4 A4 A4 A9

Ubx⁻
abd–A⁻
Abd–B⁺

3   4   4   4   4   4 10 11 12 13 14
T1 T2 T2 T2 T2 T2 T2 A5 A6 A7 A8

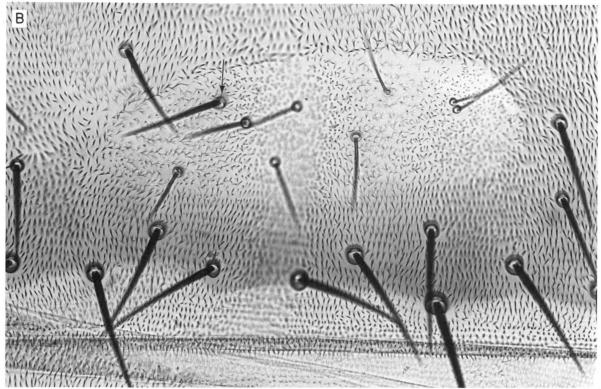

B

structures characteristic of the thorax are now found in the abdomen. The role of the *Ubx* gene is slight further back, thus parasegment 14 appears to be completely unaffected in *Ubx⁻* larvae and parasegments 12 and 13 are almost wildtype. Note that the abnormal patterns in the main part of the abdomen can be regarded as the outcome of an unusual combination of gene products, 'nonsense code words', that are not found anywhere in the body of a wildtype embryo. For the parasegments 7−9, the nonsense code word consists of the presence of the *abd-A* protein and the absence of *Ubx*. In Figure 5.3 abnormal patterns that arise from nonsense code words are indicated by red italic letters and numbers. See if you have understood the ground rules by predicting the pattern for larvae that are Ubx^- $abd\text{-}A^+$ $Abd\text{-}B^-$, and checking your answer with Figure 5.3.

Comparable results are obtained in the adult — for example, if clones of *Ubx⁻* are made in the embryo in the anteriormost compartment of parasegment 5 of the leg (which is T2p — see Box 1.2, p. 4) then the cells are transformed to T1p (which is parasegment 4). *Ubx⁻* clones in abdominal compartments posterior to A1a develop almost normally because they are *abd-A⁺* and now depend largely on that gene; however, clones of *abd-A⁻* in abdominal compartments A2a− A4a (parasegments 7−9) are transformed towards A1a (parasegment 6, Figure 5.3). If abdominal clones lack both *Ubx⁺* and *abd-A⁺* they do not appear in the anterior abdomen of the adult. The reason is presumably that the mutant cells are converted from abdominal cells into the thoracic ones of parasegment 4 — and fail to form an imaginal disc all on their own. A clone of *Abd-B⁻* cells in the dorsal epidermis of A6a (parasegment 11) will be transformed towards A4a (parasegment 9).

Perhaps because the spatial regulation of BX-C genes is so complex, dominant gain-of-function mutations are frequent; these cause adventitious expression of gene product and can be very informative. Two examples: if the *Ubx* gene is expressed in parasegment 4, the wing is transformed to a haltere (the *Haltere mimic* mutation). If *Abd-B* is misexpressed anteriorly, the A4 abdominal segment develops characteristics of A5 (the *Mcp* mutation). These types of mutations illustrate that selector genes direct the fate of cells and must, in the wildtype, be properly confined.

Figure 5.3 *(opposite)* The three genes of the bithorax complex work in combination in the wildtype. A, when genes are removed, one by one or in pairs, the parasegments are transformed as shown. Parasegment numbers are shown at the top — when the parasegments are incorrectly patterned the digits are shown in red italics. Below, the segmental numbers refer to the bands of denticles only. B, adult clone of cells that is *abd-A⁻*, the clone is located in the A3 segment and the cells are transformed and make bristles as in A1. Arrow indicates A2 bristle inside A1 clone.

The function of these genes can be removed, by mutation, and added back by heat shock constructs in which, for example, the Ubx^+ coding sequence (from a cDNA) is placed under the control of a heat shock promoter (see Box 3.2, p. 56) and transformed into flies (p. 219). If the Ubx gene is expressed at a high level everywhere, as a result of a series of heat shocks (and this can be monitored by antibody against the Ubx proteins), there is no effect on pattern in the domain where the endogenous Ubx gene is operating (parasegments 6–14). In parts of the body not under control of the BX-C, that is anterior to parasegment 5, ubiquitous Ubx expression transforms all those parts to parasegment 6 so that a cuticular pattern similar to T3p/A1a appears in all the segments of the head. These are simplifying results, because they tell us that two complicating features of Ubx expression may be overlooked for the moment at least.

First, in normal embryos the level of Ubx protein varies a great deal, making a complex pattern in each parasegment; it was natural to ascribe a function to this. However, the universal and evenly distributed expression of Ubx at high level after heat shock produces a normal cuticle pattern in the abdomen, at least in the epidermis. One exception to this delightfully simplifying story: in parasegment 5, Ubx levels are very low in most cells; after heat shock this parasegment differentiates as parasegment 6, suggesting that in this case the level of Ubx protein might be important — although several other explanations are equally plausible. If it **is** the level that counts it is an exception to the general rule that selector genes act as bistable switches.

Second, there are several forms of the Ubx protein that are the result of complex RNA splicing permutations; however it seems as if each one of these (made in different heat shock constructs which are each based on one cDNA) is equivalent in its effects on the cuticle pattern; consequently, in general, there is no need to posit different roles for the different types of transcripts in the epidermis. In other tissues there is some evidence that the alternatively spliced products do different things.

The combinatorial interactions between the three BX-C genes take various forms and include *trans* interactions that affect transcription. For example, in the wildtype embryo, the $Abd-B$ gene suppresses Ubx transcription in parasegment 13. This repression is evaded by using the heat shock construct and one can artificially introduce Ubx protein into parasegment 13. Surprisingly, these larvae are quite normal, the high level of Ubx protein has no effect on the pattern. This phenomenon has been called 'phenotypic suppression' to emphasise that $Abd-B$ products will effectively silence the Ubx protein; there must be post-translational events and interactions. Although the mechanism is not known, there are several possibilities; perhaps Ubx and $Abd-B$ protein both bind to the same or overlapping sites on the genes they regulate and *Abd-B* has a higher affinity for those sites. It would follow that, where both proteins were present, only *Abd-B* would bind and affect the transcription of these downstream genes. Observations on genes of both the BX-C and the Antennapedia complex (p. 128) suggest that complete or partial phenotypic suppression is commonplace and follows a general rule. The rule is that gene products (apart from *Deformed*), normally expressed in anterior parts, are phenotypically suppressed by proteins expressed more posteriorly. For example, the effect of high levels of *Antennapedia* protein on pattern is completely silenced by 1 dose of Ubx^+. Thus, in the wildtype, how much a gene influences the final pattern depends on which other genes are co-expressed in the same cells, in the same parasegment.

These mechanisms give the combinatorial code some properties of a hierarchy. They suggest that minor changes in pattern, by which I mean both gradual evolution through time and slight differences in the pattern of neighbouring segments, might result from changes in the binding sites of the downstream genes that give one selector gene more weight than another.

The bithorax complex has been cloned (see Box 4.1, p. 82, and The history of the bithorax complex, p. 211) and proved to be complex indeed. It is large (more than 300 kilobases) and has long introns and regulatory regions. The molecular map is shown in Figure 5.4. Each of the three genes is separate and the order given is that on the chromosome, with *Ubx* being the gene nearest to the centromere. The three transcripts are shown in black; the direction of transcription is the same for all three, from a 5' start on the right to a 3' exon on the left. Each 3' exon contains a homeobox (H) and this portion is always common to the different splicing variants of the protein. Mutations that disrupt the transcript are null alleles and some are shown in black below. The regulatory regions, shown in red above, are approximately defined by the location of regulatory mutations (named in red). In the case of *Ubx*, there is a large regulatory region in the intron. In the case of the *Abd-B* gene, there are two different forms of the protein, a longer *m* form which comes from a short transcript and a shorter *r* form that is made from a long transcript. These two forms have different patterns of expression and requirement. Both share the same homeodomain, but *Abd-B,r* is expressed only in parasegment 14 and posterior to it, and *Abd-B,m*, probably, in parasegments 10−13.

Many of the regulatory mutations (shown in red) produce either loss of the protein in **part** of its domain or gain-of-function expression in new domains (and, frequently, they do both these things). This illustrates that most elements in the long control regions ensure the proper spatial expression of the BX-C genes. It is noticeable from the map in Figure 5.4 that the order of the regions most affected by the regulatory mutations is the same order as the parts of the body. For

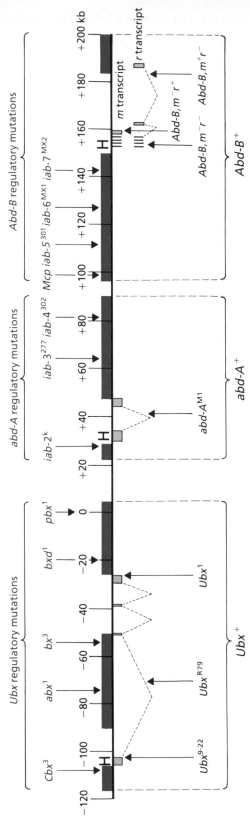

Figure 5.4 A genetic and molecular map of the BX-C. The three genes are *Ubx*, *abd-A* and *Abd-B* and their transcripts, introns and exons shown below in black. Mutations which interrupt these transcripts and/or inactivate the gene product are shown below in black. Some regulatory mutations are shown above the line in red and affect the regulatory sequences, shown as red blocks. Note the molecular map is oriented by the genes on the chromosome; so 5' to 3' is from right to left. H marks the three homeoboxes, all located in 3' exons.

example, *Cbx* (on the extreme left) affects parasegment 4; *bxd*, parasegment 6; *iab-2*, parasegment 7; *iab-4*, parasegment 9; *iab-6^{MX1}*, parasegments 10–12; and *Abd-B,r^-* (on the extreme right) affects parasegment 14.

Many studies of the pattern of expression of different selector genes have been published and some of them yield a confusing picture — confusing that is if you prefer the simple view that each selector gene is active in a specific set of parasegments. The facts are that while many selector genes do show parasegment-specific expression there are exceptions, particularly at the later stages. The borders of expression are unstable and have different limits in the dorsal and ventral parts of the embryo. The reason for this may be that the control of selector gene expression is subject to many influences, not least the *trans* effects of other selector genes and these may complicate the original pattern. Ideally, one should look at the expression of a selector gene in the complete absence of all the others and determine its primary domain; then the others could be added one by one to see how they affect that primary pattern. One example: in the ectoderm, *abd-A* is expressed from parasegment 7 (a sharp boundary exactly at the parasegment border) back to a posterior boundary within the anterior compartment of A8 (within parasegment 13). In the absence of the *Abd-B* gene, early expression of *abd-A* now extends back exactly to the posterior border of parasegment 13, showing that the posterior boundary of *abd-A* expression in the wildtype is not fixed by intrinsic properties of the gene, but rather by regulatory interactions within the BX-C. Later, in *Abd-B^-* embryos, the *abd-A* gene becomes activated in parasegment 14 and beyond, possibly due to new *trans* effects. However, the *Ubx* gene has a different intrinsic pattern; in the absence of both the *abd-A* and *Abd-B* genes, it is expressed in parasegments 5–13 but does not spread further back. More studies of this type will be needed to sort out the role of intrinsic and extrinsic factors in the spatial regulation of selector genes.

The bithorax complex has been around a very long time: the *abd-A* gene has been isolated from the locust and antibodies made against it. The expression pattern is much the same, even though locusts are such different insects from *Drosophila*. The anterior limit of expression is at the border between A1a and A1p, that is at the anterior border of parasegment 7 (exactly as in *Drosophila*), not only showing the conservation of domain but signalling the existence of compartments and parasegments in the locust. Comparison of the same gene in different species can also be made at the level of DNA sequence, and this can be very informative (see Box 5.1).

Molecular biology has not yet helped us to understand how a single protein present in a cell can change it from making part of one pattern (e.g. a wing) to making part of another (a haltere). We only know that

Box 5.1 Comparing genes and DNA sequences between species

It is very hard to find out, when one looks at, say, the 6.1 kilobases of sequence upstream of the *ftz* coding region or the 70 kilobases of intron in the *Ubx* gene, which bits are for what, which bits of the sequence have a function, which are merely spacers. One useful approach is to cut it up with restriction enzymes and place the pieces in front of a minimal promoter and reporter gene and transform flies. Another is to study mutations at the molecular level and see where sequence changes have phenotypic consequences. Another is to do footprint studies (p. 55) and identify consensus sequences — one can then use these consensus sequences to pinpoint potential binding sites elsewhere in the genome.

Still another is to compare homologous genes from different species of *Drosophila*, or even other flies or insects. This approach works because insignificant sequences tend to drift and change length, while ones that bind regulatory proteins are conserved. I believe this is a very valuable method, and that it will be increasingly used. For example, the *ftz* genes of *Drosophila melanogaster* and *D. hydei* have been compared. The *D. hydei* gene can be excised and transformed into *D. melanogaster* and it works reasonably well, showing that the basic elements of control are conserved between the two species. In *D. hydei* the whole *ftz* gene is inverted as if it has been cut out and replaced. Comparison of the regulatory regions, both 5' and 3' of the gene, reveals small, highly conserved regions interspersed in a background of sequence that bears little resemblance between the species. The lengths of these spacers also vary. Some of the conserved boxes contain sequences that homeodomain proteins recognise, the conservation indicating that the sequences are functionally important. Some such sites are not conserved, suggesting they may not be utilised.

Regulatory proteins that bind to the upstream element of *ftz* have been identified, an example is the *tramtrack* gene product. Two of the potential sites for this protein, recognised by a consensus sequence, are not conserved, and this makes one wonder about the importance of those particular sites. Comparison between species is a valuable way to identify regulatory elements which can then be tested by asking them to drive β-galactosidase expression in transformed flies. It is an 'elegant way to evade creating an endless series of deletion constructs, where mainly luck dictates how precisely the elements can be defined.'[7]

This strategy can be also used at a more basic level. Consider segmentation in the locust which, unlike *Drosophila*, occurs serially with new segments being added one by one at the posterior tip of the embryo. It is already known (from an *engrailed* antibody that recognises *engrailed* protein in many animals) that there are anterior and posterior compartments in the locust as in *Drosophila*. Analysis of the *abd-A* gene in both species shows that the selector gene recognises the same anterior parasegmental limit in both. We now need to know whether there are *ftz* and *eve* homologues in the locust — are they involved in defining parasegment borders? The answers will be illuminating in several ways; first, they will tell us whether the *ftz* and *eve* genes have a similar function in the locust — if they do the evidence will be independent and corroborate (or not!) the view of the role of pair rule genes in *Drosophila*. Second, they will indicate how far findings can be generalised from *Drosophila* to other insects and beyond. This latter is important because, at the moment, we do not know whether the genetic logic being unravelled in *Drosophila* is fundamental and general — or whether, at this level, every group has solved the problem of cellular allocation differently.

For more information, see Maier, D., Preiss, A. and Powell, J.R. (1990) *EMBO J.* **9**: 3957–3966.

Ubx is necessary and sufficient to orchestrate the differences between the two patterns. Since the *Ubx* protein is a transcription factor with a homeodomain, the expectation is that it must work through other genes, the so-called 'downstream genes of *Ubx*'. Some believe these will be a special class of regulatory genes that are specific to *Ubx* and lie between the *Ubx* master gene itself and those executive genes which either regulate or make such everyday things as cuticle proteins. I prefer the idea that there is no special class, but that the different patterns come about through *Ubx* and the other selector genes controlling a common set of regulatory genes. An example of the latter might be the *scute* gene. *scute* is an important agent in the determination of bristle pattern and is required all over the body (Chapter 7). *scute* must be differently used in the three legs because the bristle patterns differ so much, for example between T2p and T1p — a difference dependent on *Ubx*.

Could the wing and haltere depend on completely or largely distinct sets of executive genes? This looks unlikely: here are three arguments. One, two-dimensional gels suggest that wing and haltere imaginal discs have virtually all proteins in common. Two, 70 years of mutagenesis, both natural and contrived, have not produced mutations that affect, for example, cuticle structure in parasegment 6 but not parasegment 4. If there were classes of genes that responded specifically to the direction of *Ubx*, then one would expect such mutant flies. Three, when the *Ubx* and *Antennapedia* proteins are coexpressed in the same cells by placing them both under heat shock control, the cuticle pattern produced is entirely due to *Ubx*. This phenotypic suppression of *Antennapedia* is very likely due to *Ubx* winning the competition to regulate the same set of genes — for, if *Ubx* and *Antennapedia* each had its own set of subordinates, the result would surely be some mixed cuticle pattern.

Bithorax complex — the neurectoderm

One of the reasons for the complexity of the control regions of the BX-C must be the pattern of expression and function in the different germ layers. As in the epidermis, the pattern in the central nervous system is complex; some cells express *Ubx* strongly and some almost undetectably, even though they are in the same parasegmental domain. We have just seen that this is true of the epidermis also, but there this complex pattern can be obliterated by heat shock expression of *Ubx* — apparently without any consequence. So the complex pattern within parasegments of the central nervous system **may** not be important either; it is difficult to be sure because the anatomy of the central nervous system is not as well known as that of the epidermis and cuticle. In the complete absence of *Ubx*⁺, the central nervous system is transformed in exactly the same direction as the epidermis.

In the epidermis, when the *Ubx* gene is removed from a small clone of cells, they transform autonomously. There is a particular need for mosaic experiments of this kind in the central nervous system because any change in the internal organs could be due to indirect effects — for example, sensory neurones coming in from the periphery may be directly transformed by mutations in the BX-C and these could then influence the interneurones to which they connect. To find out if there is a direct effect on the central nervous system itself, that is a local requirement for *Ubx*⁺, one must remove the *Ubx* gene from the central nervous system only. Such experiments have not been done. Nevertheless the way the central nervous system derives from within the epidermis, the homeotic transformations when the BX-C is removed and the behaviour of sensory axons in genetic mosaics, all accord with the idea that the BX-C works in the central nervous system as it does in the epidermis, by acting directly on the cells of parasegments 5 and 6, and having a reduced role in the domains of *abd-A* and *Abd-B* further back.

Figure 5.5 illustrates the effect of removing parts of the BX-C on the central nervous system. In the embryonic central nervous system, one

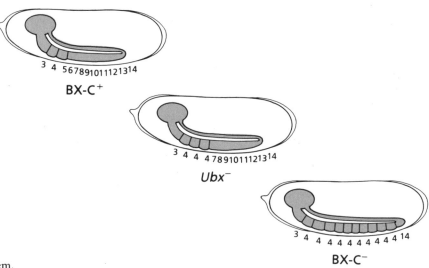

Figure 5.5 The BX-C and the embryonic central nervous system.

sees dark bands in the thoracic parasegments 3, 4 and 5 which are absent from the abdomen. In *Ubx⁻* embryos (where, in the epidermis, parasegments 5 and 6 are transformed to 4) there is an extra band at the position of parasegment 6 due, presumably, to the transformation of that parasegment to parasegment 4. In BX-C⁻ embryos, just as in the epidermis, all the 14 parasegments are transformed to the thoracic type and display characteristic bands.

The bithorax complex and *engrailed* — the somatic mesoderm

The requirement for BX-C genes in the **somatic** mesoderm is much more uncertain. The expression pattern is simpler than in the epidermis and central nervous system. In the embryo, *Ubx* is expressed in parasegments 6–12 at an even level, while *abd-A* and *Abd-B* are expressed in overlapping sets of parasegments (7–12 and, probably, 10–14, respectively) (see Figure 5.8). Cell lineage studies on the mesoderm, made both by nuclear transplantation and by mitotic recombination, indicate that at least the parasegments of the thorax are lineage units; meaning that once allocated in the embryo, mesodermal cells colonise only one parasegment. A working hypothesis is that all the mesodermal parasegments are compartments from when they are established at about the stage of gastrulation, that cells within them give descendants only within a prescribed muscle set and that these cells express a specific combination of selector genes. For example, in the embryonic somatic mesoderm, parasegment 5 expresses no *Ubx* while parasegment 6 does; coordinately, in the larva, T3 muscles have no *Ubx* antigen, while A1 muscles do. All A2 muscles have a higher level of *Ubx* antigens than their homologues in A1, consistent with the hypothesis that the two muscle sets develop independently and that the muscle parasegments are compartments. Thus, as in the ectoderm, there are reasons to believe that the mesodermal parasegments are important entities in terms of design.

How are the mesodermal parasegments founded? It seems likely that the parasegment boundaries are drawn by *ftz* and *eve*: their sharp anterior boundaries compart both ectoderm and mesoderm at the time of gastrulation and are collinear with the borders recognised by selector genes such as *abd-A*. However, the mesodermal parasegments are apparently not subdivided into anterior and posterior compartments. There are three arguments for this; two depend on the expectation that *engrailed* would work in the mesoderm as it does in the ectoderm. First, analysis of the cell lineage of the thoracic muscles revealed no anteroposterior subdivision within the segment (see Figure 4.5, p. 94). Second, although the *engrailed* gene is initially activated along the parasegment border in the mesoderm, it soon fades away. Third, there is evidence that *engrailed⁺* is dispensable in the mesoderm. This evidence consists of making adults that have mainly wildtype ectoderm cells, but with large parts of their mesoderm consisting of cells homozygous for *engrailed* lethal alleles. Such adults are normal (Figure 5.6 and Box 5.2), suggesting that the *engrailed* gene has no role in the mesoderm; even its transient expression in the early embryo is dispensable.

What determines the segment-specific pattern of muscles?

In the larva the muscle pattern is specific for different sets of segments, and in the adult the segments are even more diversified. The other

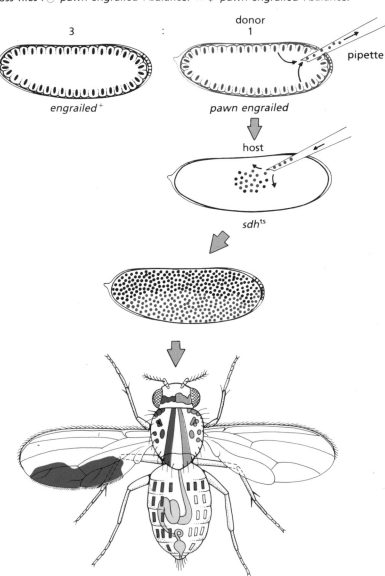

Figure 5.6 Where is the *engrailed* gene required? Mosaic flies that consist of a mixture of wildtype cells and cells homozygous for a lethal allele of *engrailed* (*engrailed*^L, shown in red) are made by transplanting nuclei (see Box 5.2).

Box 5.2 Genetic mosaics — nuclear transplantation

This is another method of making genetic mosaics and it can be very useful; the main advantage is flexibility as a greater variety of genotypes can be mixed than with other methods. Donor nuclei are taken up from a late stage 5 embryo into a needle and then transferred in small lots of about 10 nuclei into much younger, cleaving hosts. Up to 30% of the embryos that result are mosaic and many can go on, in appropriate genotypes, to make adults. If the host is $Minute^+/Minute^-$ (p. 79) and the donor $Minute^+/Minute^+$, the donor nuclei will tend to fill any compartments they enter. This technique was used to find the parasegmental allocation of some particular thoracic muscles (p. 94) as well as to investigate the genetic determination of muscle pattern (see Figure 5.7).

If the donor carries a mutant allele of a gene then one can find out in which tissues the wildtype allele of that gene is required. An example: males and females heterozygous for a cell marker (pwn^-) that is linked to a lethal mutant in the *engrailed* gene (en^L) are crossed, eggs collected and used as donors. Nuclei from these eggs are transplanted into host embryos carrying a marker mutation (sdh^{ts}). Cells that derive from the transplanted nuclei are sdh^+ and will stain blue and can be identified in the white unstained background of the sdh^{ts} host cells. Only one-quarter of the donor eggs will be pwn^-en^L/pwn^-en^L and these are identified because they give rise to mosaic offspring containing some cuticle marked with pwn^-. The *engrailed*^{L} cells were found to be defective in the ectoderm but normal in the endoderm and mesoderm. The *engrailed*^{L} allele used expresses very little antigen and is therefore probably close to a null allele; the experiment therefore suggests that the *engrailed* gene is not needed in either the mesoderm or the endoderm (Figure 5.6).

For more information, see Lawrence and Johnston (1984) (details p. 104).

derivatives of the somatic mesoderm, such as the heart, the fat body and the gonads, also develop segment-specific patterns. It is natural to assume that, as in the ectoderm, the parasegmental pattern of expression of the BX-C is directly responsible for these variations. However, the key experiment of removing a bithorax gene from the muscle cells and **from only those cells** has not been done, so even though mutants in the BX-C do affect muscle patterns, it is not clear how. For example, muscles have to attach to specific places on the cuticle; it is not surprising therefore that alterations in the cuticle pattern have been shown to affect muscle patterns. Muscles are innervated and, as we shall see below, the nerves may play an important role in determining muscle development. So the question of whether the *Ubx*, *abd-A* and *Abd-B* genes are involved directly in specifying mesoderm cells is an open one, although the pattern of their expression would certainly lead one to believe that they do.

There is one muscle that seems ideal to investigate this general question and that is a special 'strap' muscle found only in the abdominal segment A5 of adult males. In viable BX-C mutations (*Mcp*) that transform A4 into an extra A5 as well as certain *Abd-B* mutant alleles that transform A6 and A7 into adventitious A5s, both the cuticle and the muscles are transformed together — an extra strap muscle forming in the ectopic A5 segments (Figure 5.7). Mosaic flies can be made which are mixtures of mutant (*Mcp*) and wildtype tissues and all the cells marked; the points of interest are the genotype and phenotype of the cuticle, muscles and nerves in the mid abdomen. Results show that the pattern of the cuticle does not always correspond with the muscles — for example a strap (A5) muscle could attach to normal wildtype cuticle in A4. Even more surprising, the genotype of the muscle (whether male or female, *Mcp* or *Mcp*$^+$) does not always correlate with how it develops: in some cases female muscle cells even make 'male' muscles (Figure 5.7A). This suggests that non-muscle cells determine the muscle pattern. Amongst the mosaics, there were two particularly eloquent examples: in these cases, mutant *Mcp* neurones appeared to 'induce' the mesoderm of segment A4 to form a strap muscle. This happened even though the muscle cells themselves were entirely wildtype (*Mcp*$^+$) as were the cuticle and epidermis to which the muscles attached (see arrow in Figure 5.7B). The experiment implicates the neurones which innervate the muscles; it suggests that they are responsible, at least in part, for the development of specific patterns of muscles in each segment.

The determination of muscle pattern is complicated. Muscles develop in an unusual way: it seems that in *Drosophila* at least, a muscle fibre may begin life as a single pioneer or founder cell which becomes identified in the mesodermal epithelium of the embryo. Possibly the mechanisms that single these pioneer mesoderm cells out from the remainder share features with those that single out neuroblasts from the epidermis (Chapter 7). Nerves cannot be directly involved in the specification or selection of these pioneer cells for, in the *Drosophila* embryo, they develop **after** the ground plan of the muscles is already visible as a pattern of small syncytial fibres. Once the pioneer cells are chosen they are joined by 'secondary' mesodermal cells which fuse with them to form syncytial muscle fibres. The muscle fibres, or at least one end of them, migrate across the epithelial surface to take up their positions, where they then attach to the cuticle. The route they migrate along and the points they attach to are written in the epidermis — this has been shown by clever grafting experiments (in a beetle, *Tenebrio*) where the muscle pattern can be altered by transplanting small pieces of epidermis, the 'roads' on which the muscles walk. This information, together with the experiments implicating the neurones just described, tells us that the muscle pattern is the outcome of several different kinds of interactions between mesoderm and ectoderm.

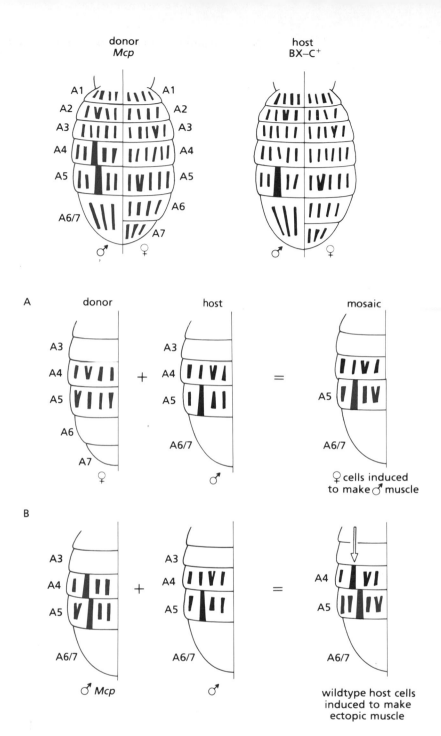

Figure 5.7 What determines the pattern of muscles? Flies that are a mixture of *Mcp* (red) and wildtype (black) cells are made by nuclear transplantation. Arrow marks a male muscle which, although made by wildtype cells, forms a muscle in A4 that is always found in *Mcp* flies (but never in wildtype flies).

There are also indications that the muscle pattern is dependent on genes which work locally within specific muscles. We have seen that the *Ubx* gene is expressed in certain parasegments of the mesoderm and it would seem likely, if unproved, that the gene is partly responsible for the pattern of those parasegments. In the epidermis, other homeobox genes (such as *engrailed*, *gooseberry*) are expressed in only parts of the parasegment, and in the somatic mesoderm there are now examples of genes expressed in only subsets of the muscles (S59, p. 95). It is not yet known what the roles of these genes are, as mutants have not been made, but it is a good guess that they are responsible for specifying the state of the cells in just those muscles.

The bithorax complex in the visceral mesoderm

We have discussed why the visceral mesoderm, which generates the smooth muscle sheath around the gut, should be regarded as a separate germ layer. A main argument is that the expression pattern of the genes of BX-C is different from that in the somatic mesoderm. *Ubx* is expressed in parasegment 7 only, *abd-A* from parasegment 8 to 12 and *Abd-B*, 11 to 14 (Figure 5.8).

There are some hints that the *Ubx* and *Antp* genes may have a local role in the visceral mesoderm, although genetic mosaics have not been made to test the idea rigorously. This role is concerned with the morphogenesis of the gut as a whole, including the innermost tube of endodermal cells. In *Ubx*⁻ embryos, the gut does not develop properly; in particular it lacks the second constriction that forms near parasegment 7 (Figure 5.8). In these *Ubx*⁻ embryos, the *decapentaplegic* gene product (see p. 66) is nearly absent in the visceral mesoderm and also local expression of the homeobox-containing gene *labial* is missing in the underlying endoderm cells. Most tellingly, in *abd-A*⁻ embryos, *Ubx* expression extends back all the way from parasegment 7 to 12; concomitantly with this, *decapentaplegic* and *labial* expression also extend back to parasegment 12. It follows that there is a hierarchy in the action of these genes; it is clear that *Ubx* is upstream of *decapentaplegic*, which itself is required for *labial* expression or, to put it another way, the local expression of *Ubx* in the visceral mesoderm patterns the expression of *labial* in the gut, and it is the local expression of *labial* that is more directly responsible for the constriction. As expected, in *labial*⁻ embryos the constriction does not form. Because the *decapentaplegic* gene encodes a secreted factor, there is reason to suggest that this is the messenger carrying some inductive signal from the visceral mesoderm to the gut (Figure 5.9). This might appear a baroque way of patterning the gut. But it makes sense, for the endoderm is not segmented and therefore, if the gut is to have its parts differentiated, why not make use of the ready-made map written in the visceral mesoderm as zones of BX-C proteins?

These studies of the BX-C in the four main germ layers yield a consistent picture: in the ectoderm, complex patterning specifies epidermis and neuroblasts. A simple pattern in the somatic mesoderm

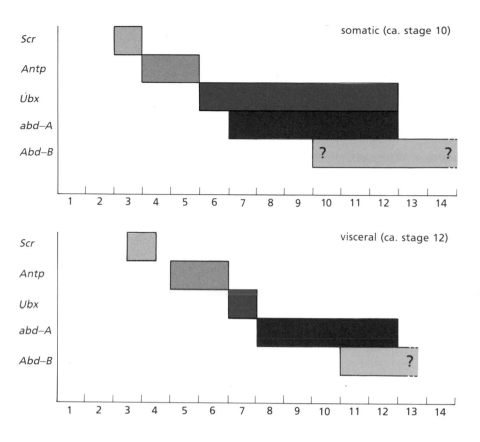

Figure 5.8 Patterns of expression of selector genes in the embryo differ in the somatic and visceral mesoderm — numbers refer to parasegments. Note, in the visceral mesoderm, *Sex combs reduced* (*Scr*) is not expressed in a parasegmental register. Below, a sketch of embryonic gut shows the pattern of gene expression in the visceral mesoderm as it relates to the structure of the gut (the caeca and three constrictions).

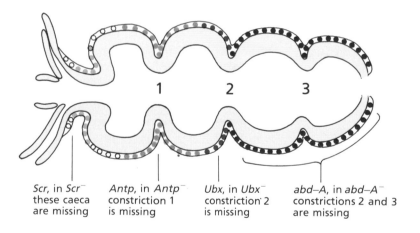

Scr, in *Scr⁻* these caeca are missing

Antp, in *Antp⁻* constriction 1 is missing

Ubx, in *Ubx⁻* constriction 2 is missing

abd–A, in *abd–A⁻* constrictions 2 and 3 are missing

is elaborated by the use of specific interactions between the inducing epidermis and responding mesoderm. The simplest pattern of BX-C gene expression is found in the visceral mesoderm and some aspects of this are transferred to the endoderm to sponsor diversification of parts of the gut. The whole picture rings of evolution, it illustrates that different germ layers can evolve independently as controlling elements

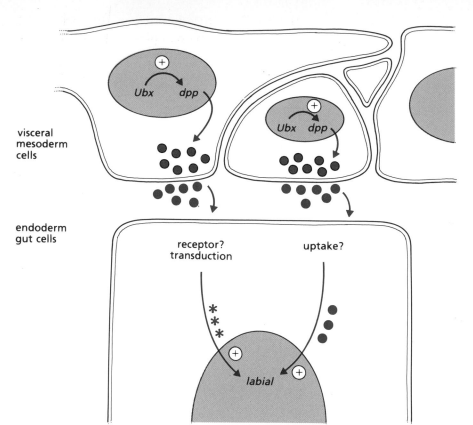

visceral
mesoderm
cells

endoderm
gut cells

receptor?
transduction

uptake?

labial

Figure 5.9 Model of induction between two germ layers; the visceral mesoderm and the endoderm. Experiments suggest that *Ubx* expression in parasegment 7 of the visceral mesoderm lies upstream of the *decapentaplegic* (*dpp*) gene, whose product is secreted and signals to the endoderm, a signal leading to local activation of the *labial* gene.

are added to a common set of regulatory genes. The endoderm is imagined to be inherited, rather little changed, from ancestors that were entirely unsegmented. Just to emphasise the point that genes accumulate different controlling elements in evolution, the control of *Ubx* expression in the visceral mesoderm is special — the activation of *Ubx* there is autoregulatory, meaning that expression depends in some way on the *Ubx* protein itself.

The Antennapedia complex

The Antennapedia complex, ANT-C, is a genetic system that shares many features with the BX-C. It contains homeobox genes that determine the identity of the parasegments in front of parasegment 5, where the *Ubx* gene has its most anterior realm of action. The best known genes are *Deformed* (a role in the ectoderm of parasegments 0 and 1), *Sex combs reduced* (*Scr*) (2 and 3) and *Antennapedia* (4 and 5). The genes are clustered together, but also interspersed with other genes, such as *ftz* and *zerknüllt*, which are different in nature although they also contain homeoboxes. Since, in the beetle *Tribolium*, the ANT-C and the BX-C are together at one chromosomal location, it seems certain that the two gene complexes are related and probably derive from the serial duplication of an original gene. In the absence of genes of the ANT-C, parts of the body in the embryo, or of specific organs in the adult, are transformed in a homeotic way. For example, in Scr⁻ embryos, parasegment 3 is transformed towards parasegment 4 and clones of Scr⁻ cells in the adult T1 leg give patterns appropriate to T2. *Antennapedia* is expressed in parasegment 4 and coexpressed with *Ubx* in parasegment 5 (T3a): if *Ubx* is removed from T3a it is transformed into T2a. If *Antennapedia* is removed from cells in the adult mesothorax (T2a), antennal structures are made. If **both** *Ubx* and *Antennapedia* are removed from cells in T3a, they make antenna. The order in which the genes act is important; the *Ubx* gene is needed early in development to confine function of the *Scr* gene to its proper domain. It is as if all these different selector genes, through acting on each other as well as on target genes, combine to specify the patterns made. The genes of the ANT-C, like those of the BX-C, are expressed throughout development and in both complexes the pattern of expression differs in the germ layers (see Figure 5.8).

Spatial regulation of selector genes

How do the various genes of the BX-C and the ANT-C come on in the appropriate regions? This question has two parts: first, how is it that *Ubx*, for example, is activated in the middle of the embryo backwards and, second, how is it that the boundaries of *Ubx* expression at its limits coincide exactly with parasegment boundaries (parasegment 5 in the epidermis, parasegment 6 in the somatic mesoderm and parasegment 7 in the visceral mesoderm)? The answers to these two questions are unknown, we can only speculate.

Experiments on hierarchy show that the positional information generated by the *bicoid* gradient and the gap genes such as *hunchback* and *Krüppel* are upstream of *Ubx* — if they are altered, *Ubx* expression is moved. The gap genes might regulate the genes of the BX-C and ANT-C directly. Alternatively, they might first render them 'open-for-business' in limited regions of the embryo; they could change the structure of the gene in some way or bind a factor to the upstream region of the gene. The image is of a structural change in the chromatin which would willy-nilly affect any enhancers and control elements transposed into the region, so that they too would become open-for-business. Open-for-business would mean that the DNA could accept *trans* acting factors, such as other homeobox genes, or more local cellular controls.

As regards the precise limits to *Ubx* expression at the cellular level, it is likely that the *Ubx* gene responds to the parasegment borders as laid down by the anterior margins of *ftz* and *eve* (p. 97). This might be the only way to get the set of cells that express a selector gene to be exactly the same set that belong to a particular parasegment by lineage. Given the noise in the system, as evidenced by the variable wiggle in the nascent parasegment border, and given the importance that every cell in a parasegment should start development as part of the same compartment with the same selector genes active, such cell-by-cell correspondence is vital.

We can look at this matter in the visceral mesoderm. For example, *Antennapedia* is expressed in parasegments 5 and 6, *Ubx* in parasegment 7 and *abd-A* in 8−12. This pattern is maintained by mutual suppression; if the *abd-A* gene is removed, *Ubx* expression spreads back into its territory. In *eve⁻* embryos, where no parasegment borders are formed, *Antennapedia*, *Ubx* and *abd-A* are not expressed. In *ftz⁻* embryos, where, as we have seen (p. 99) the anterior borders of odd-numbered parasegments are not affected, *Antennapedia* expression is normal. However, the normal limit of *abd-A* expression is the anterior border of parasegment 8 and this border is missing in *ftz⁻* embryos. This creates a problem for *abd-A*, which now stops at the extant border of parasegment 9. This leaves the cells that would normally make parasegment 8 unoccupied by *abd-A*, and (as in *abd-A⁻*) the expression of *Ubx* extends further back. The result, observed in *ftz⁻* embryos, is a *Ubx* band of expression filling the parasegment 7 and 8 territories and the backwards displacement of the *abd-A* domain. Parasegment borders may be necessary to delimit expression of the homeotic genes, both in the visceral mesoderm and in the other germ layers.

Other selector genes?

Although *engrailed*, the BX-C and the ANT-C are exemplary selector genes, there may well be others. Remember, the original picture of a selector gene's role was to specify pattern in a compartment or set of compartments. Mutations in such genes should cause changes in patterns which may be completely or partially homeotic; the definition does not demand that the proteins contain a homeobox, although one would expect them to influence the transcription of other genes by binding to regulatory regions of the DNA. One possible example is the *fork head* gene, which is expressed in the terminal regions of the embryo in those parts that will make the foregut, hindgut and midgut. In *fork head⁻*, the invaginating gut is missing and the cells at both ends make head structures instead. The expression pattern is very precisely confined to just those cells that invaginate, which suggests that the expression of this gene (like other selector genes such as *engrailed* and *Ubx*) may **determine** the cell state, the capacity to invaginate and make gut. As with those genes *fork head* is expressed at

Box 5.3 Mutagenesis

Mutagenesis has become, increasingly, a key part of the geneticists' strategy. In the very early days workers relied mainly on mutations that turned up spontaneously, later they began to make them. Nowadays it is usual for mutagenesis experiments to have a precise aim — such as causing lesions in a particular gene, affecting genes in a particular biochemical or developmental pathway, or collecting mutations in a defined region of a chromosome. The mutations themselves are usually produced by feeding the male flies with a chemical mutagen or treating them with some kind of damaging irradiation such as X-rays. A typical hit rate for a gene is one in 2000–10 000 mutagenised sperm.

It is most convenient to devise an F1 screen, meaning that new mutations can be collected in viable fertile flies which are themselves the immediate offspring (F1) of the mutagenised males. This can be a problem as many mutations are lethal and two such mutations *in trans* mean no fly to collect and breed from. There are many ways around this problem: for example, in order to collect new mutations at the *engrailed* locus *engrailed¹* was used, a viable allele that, when *in trans* to a lethal mutation or even a deficiency for the *engrailed* gene, gave fertile flies with a recognisable wing phenotype. It was therefore a simple task to cross mutagenised males to *engrailed¹* flies and screen the progeny for the wing phenotype. If no such convenient mutation exists one can sometimes be made by starting with a lethal mutation, generating a weak allele that gives viable flies with some easily detectable phenotype and then, in a second step, collect new lethal mutations *in trans* to the weak allele.

Another standard approach is to take a dominant gain-of-function allele and mutagenise flies carrying it, screening for revertants to wildtype. Such revertants usually knock out the gene (giving deletions or null alleles) or, sometimes, prove to be dominant suppressors of the phenotype. Both types of mutations can be useful.

If an F1 method cannot be devised then the chromosomes that may carry the mutations must be collected first and each amplified and tested in a single breeding tube. There, a new lethal would result in a missing class of flies in a single tube and the mutant chromosome can be collected from surviving siblings. This method was used to collect the lethal mutations in the bithorax complex (p. 211).

It is beginning to look as if some genes are being missed — these are ones where there is redundancy, where two genes have a similar overlapping function and where one can be dispensed with. Mutations in one of the genes will not show up as the other can make do. An example is the two *gooseberry* genes which were only found by means of a deficiency that takes out both. This presents scientists with a problem which is difficult to solve by mutagenesis to recessive loss-of-function mutations. Dominant gain-of-function mutations can still be found, however.

Mutagenesis is being increasingly used to dissect and extend genetic pathways. It often happens that a mutation m^- is recessive, m^-/m^+ giving a wildtype fly. However a single dose of another recessive mutation n^-, if it affects a process linked functionally to the role of m^+, may reduce the efficiency of the m^+ allele. The result may be that the double heterozygote

m^+/m^-; n^+/n^- may show some phenotype. An explanation: if the n^+ function is to make a substrate for the m^+ gene product to work on, there may be less substrate in n^+/n^- embryos. In this case the m^+ gene itself, now in reduced amount, may be unable to process enough of the substrate and there will be significant shortfall of product. Thus if an F1 screen is done in m^+/m^- flies the n gene could be identified (or vice versa). Using this strategy genes that are functionally linked can be collected.

One more example: consider a gene a that is expressed in particular parts of the body. Now make transformed flies in which this gene is artificially placed under the control of a promoter from gene b and is therefore expressed in some extra cells, resulting in a dominant phenotype. Mutagenise the transformed flies and select flies that show changes in this dominant phenotype. Apart from damage to the transformed construct itself, selected flies will carry mutations of two kinds. First, there will be mutations in genes which affect the control of promoter b, these genes are upstream of b. Second, there will be mutations in genes that are under the control of gene a, these genes are downstream of a. To do this effectively one needs a dominant phenotype in which subtle alterations can be easily detected — one is looking for dominant effects of mutations in the target genes and these may be slight.

For more information, see Ashburner (1989) (details p. 22).

least until the end of embryonic development. *fork head* encodes a protein which is probably metal binding. It is localised in the nucleus and is predicted to be yet another kind of transcription factor. Since the formation of gut may well be very old in evolution, and may well have preceded the development of segments, it may not be so surprising that the gene should be so different from those of the BX-C, for example.

Male and female flies are different in many ways; in size, shape, in patterns and in the specific proteins expressed. All these differences are under the control of one genetic system which initially reads the ratio of X chromosomes to autosomes and ultimately translates this into the two versions of embryos, larvae and flies. The final gene in this pathway is called *doublesex*, which makes different products in the male and the female. Just as *Ubx* selects the pathway of development in a region of the body, so these gene products appear to select a pathway of development in the whole. Therefore, *doublesex* is a type of selector gene. It is not yet known whether all cells in the male and female differ or whether there are parts (like the endodermal midgut?) which are asexual.

No doubt there are other selector genes to be discovered, and one way of doing this is to collect homeobox genes. An example: *empty spiracles* is a gene isolated in two ways first as a mutant with defects in the developing head and second on a Southern blot as a piece of DNA containing a homeobox. Analysis has shown that *empty spiracles* is expressed anterior to the *Deformed* stripe in the blastoderm; it is not yet clear whether it fits the definition of a selector gene. Another candidate is *Distal-less* which encodes a homeodomain protein and is expressed in those cells which will form appendages. In *Distal-less*⁻ embryos the vestigial appendages, such as the Keilin's organs and mouthpart sensilla, do not form, showing that the expression of the gene is an essential component for appendages and suggesting, again, that the protein is responsible for giving the cells a specific identity or fate. However it is not known whether it labels a compartmental set of cells.

The outcome of transiently expressed genes that are primarily concerned with generating positional information (*bicoid*, *Krüppel*, *even-skipped*) is the correctly located activation of selector genes (*engrailed*, *Ubx*). Selector genes remain expressed for long periods, they determine cell states and select developmental pathways; they are all transcription factors. We now know something about how selector genes become activated in the right place, but very little about how they produce the patterns they are responsible for.

Further reading

REVIEWS

Affolter, M., Schier, A. and Gehring, W.J. (1990) Homeodomain proteins and the regulation of gene expression. *Curr. Opin. Cell Biol.* **2**: 485–495.
Duncan, I. (1987) The bithorax complex. *Ann. Rev. Genet.* **21**: 285–319.
Garcia-Bellido, A. (1975) Genetic control of wing disk development in *Drosophila. Ciba Found. Symp.* **29**: Cell Patterning, pp. 161–182.
Hodgkin, J. (1990) Sex determination compared in *Drosophila* and *Caenorhabditis. Nature* **344**: 721–728.
Kaufman, T.C., Seeger, M.A. and Olsen, G. (1990) Molecular and genetic organization of the Antennapedia gene complex of *Drosophila melanogaster. Adv. Genet.* **27**: 309–362.
Lewis, E.B. (1978) A gene complex controlling segmentation in *Drosophila. Nature* **276**: 565–570.
Peifer, M., Karch, F. and Bender, W. (1987) The bithorax complex: control of segmental identity. *Genes Dev.* **1**: 891–898.
Scott, M.P., Tamkun, J.W. and Hartzell, G.W. (1989) The structure and function of the homeodomain. *BBA Rev. Cancer* **989**: 25–48.

SELECTED PAPERS

BX-C in the epidermis

Bender, W., Akam, M., Karch, F., Beachy, P.A., Peifer, M., Spierer, P., Lewis, E.B. and Hogness, D.S. (1983) Molecular genetics of the bithorax complex in *Drosophila melanogaster. Science* **221**: 23–29.
Casanova, J., Sánchez-Herrero, E. and Morata, G. (1986) Identification and characterization of a parasegment specific regulatory element of the *Abdominal-B* gene of *Drosophila. Cell* **47**: 627–636.

González-Reyes, A. and Morata, G. (1990) The developmental effect of overexpressing a *Ubx* product in *Drosophila* embryos is dependent on its interactions with other homeotic products. *Cell* **61**: 515−522.

Gould, A.P., Brookman, J.J., Strutt, D.I. and White, R.A.H. (1990) Targets of homeotic gene control in *Drosophila*. *Nature* **348**: 308−312.

Morata, G. and Garcia-Bellido, A. (1976) Developmental analysis of some mutants of the bithorax system of *Drosophila*. *Wilhelm Roux's Archives* **179**: 125−143.

Sánchez-Herrero, E., Vernos, I., Marco, R. and Morata, G. (1985) Genetic organization of the *Drosophila* Bithorax complex. *Nature* **313**: 108−113.

Tear, G., Akam, M. and Martinez-Arias, A. (1990) Isolation of an *abdominal-A* gene from the locust *Schistocerca gregaria* and its expression during early embryogenesis. *Development* **110**: 915−925.

BX-C in the mesoderm

Lawrence, P.A. and Johnston, P. (1986) The muscle pattern of a segment of *Drosophila* may be determined by neurons and not by contributing myoblasts. *Cell* **45**: 505−513.

BX-C in the nervous system

Jiminez, F. and Campos−Ortega, J.A. (1981) A cell arrangement specific to thoracic ganglia in the central nervous system of the *Drosophila* embryo: its behaviour in homeotic mutants. *Wilhelm Roux's Archives* **190**: 370−373.

Distal-less

Cohen, S.M. (1990) Specification of limb development in the *Drosophila* embryo by positional cues from segmentation genes. *Nature* **343**: 173−177.

empty spiracles

Dalton, D., Chadwick, R. and McGinnis, W. (1989) Expression and embryonic function of *empty spiracles*: a *Drosophila* homeobox gene with two patterning functions on the anterior-posterior axis of the embryo. *Genes Dev.* **3**: 1940−1956.

fork head

Weigel, D., Jürgens, G., Küttner, F., Seifert, E. and Jäckle, H. (1989) The homeotic gene *fork head* encodes a nuclear protein and is expressed in the terminal regions of the *Drosophila* embryo. *Cell* **57**: 645−658.

Gap genes act upstream of selector genes

Harding, K. and Levine, M. (1988) Gap genes define the limits of Antennapedia and Bithorax gene expression during early development in *Drosophila*. *EMBO J.* **7**: 205−214.

Homeobox genes in other animals

Chisaka, O. and Capecchi, M.R. (1991) Regionally restricted developmental defects resulting from targeted disruption of the mouse homeobox gene *hox-1.5*. *Nature* **350**: 473−479.

Stuart, J.J., Brown, S.J., Beeman, R.W. and Denell, R.E. (1991) A deficiency of the homeotic complex of the beetle *Tribolium*. *Nature* **350**: 72−74.

Visceral mesoderm

Immergluck, K., Lawrence, P.A. and Bienz, M. (1990) Induction across germ layers in *Drosophila* mediated by a genetic cascade. *Cell* **62**: 261−268.

Tremml, G. and Bienz, M. (1989) Homeotic gene expression in the visceral mesoderm of *Drosophila* embryos. *EMBO J.* **8**: 2677−2685.

SOURCES OF FIGURES

For details, see above.

Figure 5.1 See Duncan (1987), Lewis (1978) and Peifer *et al.* (1987).

Figure 5.2 See Morata and Garcia-Bellido (1976).

Figure 5.3 See Bender *et al.* (1983), Casanova *et al.* (1986), González-Reyes and Morata (1990) and Sánchez-Herrero *et al.* (1985). Photograph courtesy of G. Morata.

Figure 5.4 See Bender *et al.* (1983), Duncan (1987), Lewis (1978), Peifer *et al.* (1987) and Scott *et al.* (1989).

Figure 5.5 After Jiminez and Campos-Ortega (1981).

Figure 5.6 See Lawrence and Johnston (1984) (details p. 104).

Figure 5.7 See Lawrence and Johnston (1986).

Figure 5.8 See Tremml and Bienz (1989).

Figure 5.9 See Immergluck *et al.* (1990).

Plate 5.1 Photograph courtesy of E.B. Lewis.

6 Positional information and polarity

'The molecular mechanism for large scale patterning ... is unknown ... answers will be found by scientists prepared to think and work unconventionally.' [9]

THE GROUPS OF CELLS co-allocated to the same developmental pathway have to work together to produce a piece of the body of the correct shape, size and pattern. Individual cells are oriented in the plane of the epithelium and they must act according to their position in the group; both these features depend in part on gradients of positional information. Competition between cells is an important part of controlling growth.

Pattern formation, gradients and polarity

Once the cells have been allocated to parasegments, and once they have been given a genetic address, in the sense that selector genes such as those of the BX-C tell them which part of the body pattern they will be making, new questions arise: how does each cell know which part of the pattern it is to make? How is the polarity of each cell determined? How is growth controlled so that scale and proportion are achieved? These questions can be best approached by studying the later stages of development, such as occurs in the imaginal discs of flies, or in the postembryonic growth of other insects.

The analysis of cell lineage shows that there is great flexibility built into the process of growth. For example, cells in the developing wing divide stochastically, even sister cells may divide different numbers of times. The excessive size of $Minute^+/Minute^+$ clones in a $Minute^+/Minute^-$ background illustrates this flexibility, for one cell which would normally generate some 5% of the anterior wing compartment now makes nearly 100% and the other cells give way — yet the wing ends up exactly the right size and shape. The growth of embryos is very robust; if more than half of all cells are killed at random by X-rays, some individuals can still survive; this must mean that no cells, or very few, are indispensable. These results suggest that cells read and respond to their position flexibly; the mechanism is not one of acting out a rigid programme. So how is it done?

I believe the answer lies in experiments on other insects, such as moths and bugs (Hemiptera), and because I think the answer will be universal, it will apply to flies too.

Cell polarity

In what follows, cell polarity plays a large part. Cells are polarised in the plane of the epithelium and this fact receives little attention from embryologists, perhaps because cells in vertebrates are generally coy about revealing their polarity. Insects are particularly helpful here because the cuticle secreted by the epidermis is usually anisotropic — there are chitin fibres that are laid down along the body axis and

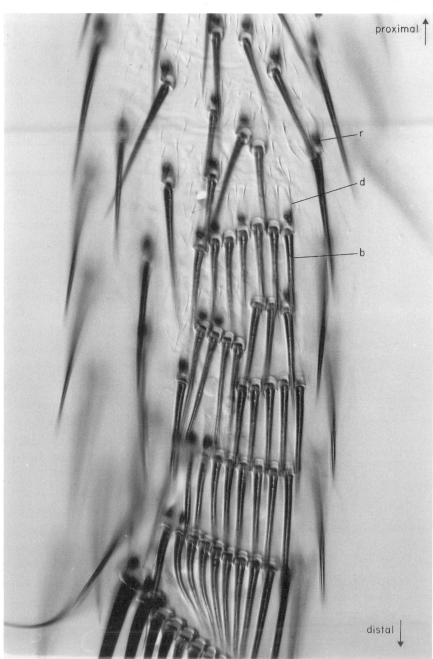

Figure 6.1 Part of the leg to show polarity. Note that the small and fine cell-hairs or denticles (d), as well as the bristles (b) point distally. The bristles form mediolaterally oriented rows. There are also bracts (r).

details of sculpturing of the surface are oriented. The leg of *Drosophila* provides a good example (Figure 6.1). The epidermal cells secrete oriented denticles which project like arrows from the surface, as do the bristles which originate as cellular outgrowths containing bundles of micro-tubules. Each bract, which comes from an epidermal cell just **proximal** to the base of a bristle, also points distally. Finally there is the **sequence** of pattern elements, the segments of the limb (coxa, trochanter, femur, tibia, tarsi) which imply a polarity in the pattern. An example from vertebrates is the beat of cilia, which is strongly polarised with an oriented active sweep: cilia are found in many places from amphibian embryos to the trachea of humans. I suspect all epithelial cells in an embryo have information of both their position in an organ and their orientation within it; this 'positional information' is used to determine their differentiation and the orientation of any polarised structure.

Gradient experiments

In the larvae of the blood-sucking bug *Rhodnius* it is easy to transfer squares of cuticle with epidermis attached from one place to another or from one individual to another. On the surface of the adult cuticle is a simple pattern of ripples which act as a record of the cells' polarity at the time the cuticle is secreted. If a piece of larval epidermis is simply taken out and replaced or transferred from one segment to an equivalent place in another, the adult cuticle is undisturbed. This is even so when the epidermis is transferred along the mediolateral axis (Figure 6.2). But if the epidermis is moved up or down the anteroposterior axis there is discordance between the graft and host and the result is changing cell polarities and altered ripple patterns. If the piece is rotated through 180° and then replaced, the adult cuticle shows a curious pattern of two whorls (Figure 6.2). Similar experiments were done on a moth larva, and the result looked at in the adult scales; the scales took up a rotated orientation in the middle of the grafted piece of integument but at the edges they were disposed in intermediate orientations as if, like compass needles or iron filings, they were responding to fields emanating from both host and graft. A related phenomenon is found in the plant-feeding bug *Oncopeltus*. Some individuals have a gap in the segment boundary and associated with it there is a rather precise pattern of hairs with altered orientations (Figure 6.3). All these results can be modelled in terms of a segmental gradient that is responsible for the polarity of the cells and provides them with positional information telling them their location within the anteroposterior axis of the segment.

The model is as follows: a dynamic gradient of some scalar variable, which could be, but might not be, a concentration gradient of some substance or 'morphogen', is established between the boundaries of the segment. The vector (that is, an arrow that points down the steepest

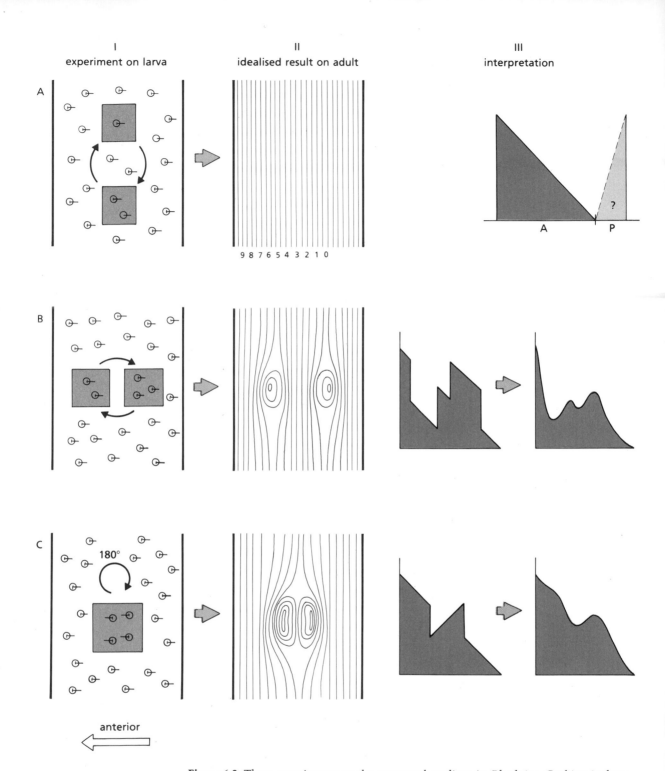

Figure 6.2 Three experiments on the segmental gradient in *Rhodnius*. Grafting is done on the larva (column I) and the operation studied in the adult. The gradient in the small posterior compartment (P) has not been studied and is represented, provisionally, by the steeply rising dashed line.

139 POSITIONAL INFORMATION AND POLARITY

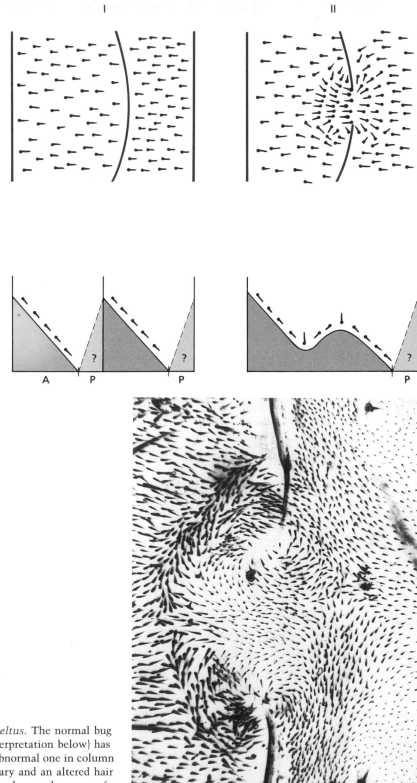

Figure 6.3 Hair patterns in *Oncopeltus*. The normal bug in column I (with the gradient interpretation below) has an intact segment boundary. The abnormal one in column II has a gap in the segment boundary and an altered hair pattern. The photograph shows the abnormal pattern of hairs.

slope at each point) determines the polarity of the cell at that point and the scalar (e.g. the concentration of the molecule) gives the position in the anteroposterior axis. We have already discussed examples of morphogen gradients (*bicoid*, p. 30; the overlapping *hunchback* gradient, p. 58) but this must be somewhat different since it operates not in a syncytium, but in a sheet of cells. If a piece of the gradient is rotated, cells at different levels of the scalar are juxtaposed (e.g. 4 and 1) and there is a blending of the levels (to give 3 and 2) by a process equivalent to diffusion of the molecule between the apposed cells. In the case of the 180° rotation of squares of integument in *Rhodnius*, the ripples act like contours in a map of mountains and the two whorls represent a peak and a valley (Figure 6.2). Similarly, in the case of the moth, the scales indicate the vectors, pointing down the local slopes. Again, in *Oncopeltus*, the gap in the segment boundary allows the juxtaposition of the posteriormost and anteriormost cells of the segment resulting in blending of the levels and new orientations of the bristles (Figure 6.3).

Each contour follows a line of cells at a similar level of positional information. In the wax moth *Galleria*, there is a ridge which runs across the segment at a particular level, and rotation of a larval square including this ridge gives an isolated ellipse and a deflected entire ridge in the adult. The gradient model predicts the result (Figure 6.4). This experiment gives strong testimony for some kind of gradient, for what other model could predict such a strange result?

Another example: the bug *Dysdercus* looks like a football jersey, each segment has cells that are alternatively red (anterior half of segment) and white (posterior half of segment). There are mutants which lack the enzymes to make the red pigment, and, in these, the anterior half of the segment is distinct but grey in colour (Figure 6.5). If a piece of epidermis from the white region of a wildtype *Dysdercus* is grafted into the grey region of a mutant host, there is a blending interaction between the cells at different levels in the gradient. The result is that along the edge of the graft the wildtype cells are brought up to the level appropriate for making red pigment, and red pigment is synthesised (Figure 6.5A). This only happens when the grafted cells are in contact with more anterior cells, not when they contact host cells in the posterior white zone. In Figure 6.5B the graft is moved from wildtype to mutant to a corresponding position. The operation kills the cells around the edge of the graft and wound healing occurs as grafted cells enter the territory where they should be red, and those graft cells, but no other, become strongly red. Clearly, therefore, the outcome of interaction between mutant host and wildtype graft is dependent on position.

When the host and graft cells that are apposed are from the same position in the anteroposterior axis little happens, apart from wound

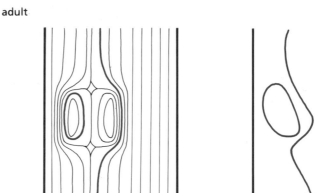

operation on larva

180°

presumptive ridge

before rotation

after rotation

adult

predicted result after 'diffusion'

actual result

Figure 6.4 180° rotation experiment on *Galleria*. The model is that cells along one particular contour in the gradient landscape differentiate as a ridge (shown in red).

healing. When they are from different positions, interaction leads to the formation of cells of the type normally found intervening between the host and graft cells. This interaction is often associated with extra cell divisions and for this reason the process has also been called intercalation. It should be remembered however that intercalation is the **outcome** of an interaction and is not the same thing as the interaction itself. The specification of the positional values of the intervening cells has to be explained, and one of the best ways to model that specification (the cell types that are formed, the way they are ordered in space and their polarity) is to resort to the gradient model. Even the gradient model is far from a mechanism, but does direct the mind towards certain mechanistic explanations.

The simple gradient model is too abstract however, the real situation is complex; not only are there extra cell divisions, cells may migrate — for example, after 90° rotation of squares of epidermis the grafts have been observed to rotate, probably due to cell migration at the edges with the middle being carried passively around. Also, grafts that are

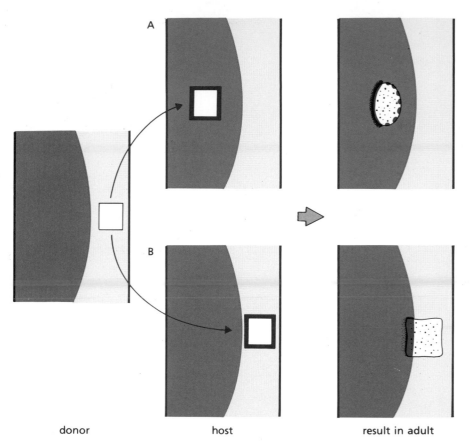

Figure 6.5 Two grafting experiments in *Dysdercus*. When grafted cells interact with anterior, but not posterior, host cells they become more anterior in character and make red pigment. The donors are male and the hosts female; only male tissue makes glands in the adult (marked by dots in the grafts).

donor   host   result in adult

transposed in the anteroposterior axis round up while control grafts do not, suggesting that there are differences of cell affinity along the axis of the segment. These factors will influence the patterns formed after transplantations. Nevertheless, whatever the details, the experiments illustrated in Figures 6.2–6.5 show that a gradient of positional information in the segmental axis is the key determinant of differentiation and polarity.

A detailed picture of the gradient landscape can be gleaned from looking carefully at polarity and its relation to differentiation. In one experiment on the wax moth *Galleria*, a piece of epidermis was transplanted from near the segment margin into the main part of the segment of a caterpillar (Figure 6.6). There are seven different types of cuticle that repeat from segment to segment and the nuclei of male and female *Galleria* are distinct, allowing marking of host and graft cells. In the larva, a small square that would have made region 5 was transplanted into region 3. In the adult, all the cells of the donor made cuticle of type 5 as expected, but, as a result of interaction between the cells of graft and host, many host cells near the graft edge were transformed and made scales of type 4 (shaded nuclei). Figure 6.6 also illustrates the effects on the polarity of these cells as indicated by the orientation of

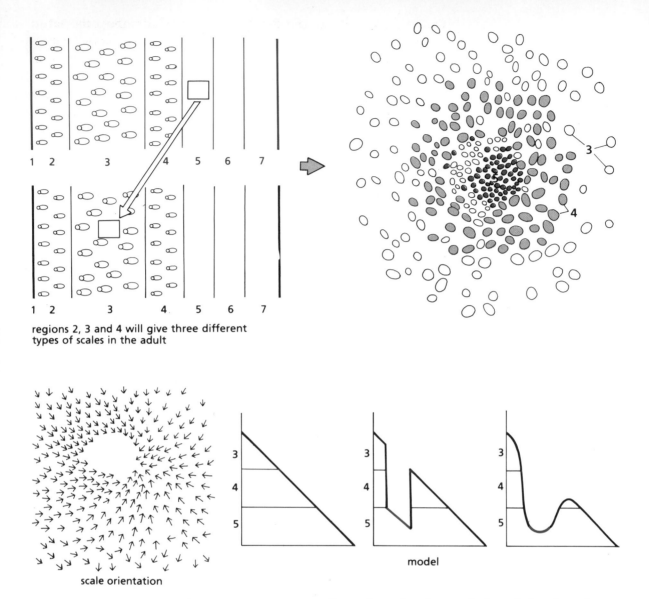

regions 2, 3 and 4 will give three different
types of scales in the adult

scale orientation

model

Figure 6.6 Experiment on *Galleria*. The graft is done in the larva, before the scales
develop. The grafted nuclei (red) can be distinguished from those of the host — the graft
induces cells of the host to produce scales of type 4 (shaded). The orientation of the
scales is shown by the arrows, which point towards the graft — gradient interpretation
shown below right.

scales; they pointed down the gradient, like vectors pointing into the
'pit' in the gradient landscape. These results illustrate the close relationship between differentiation and cell polarity. Notice that in this case
all the changes occurred in host cells; the blending was not evenhanded.
One explanation for this might be that the grafted segment border cells
are dominant and fixed in their positional values, imposing a gradient
on the more amenable cells in between.

Quantitative analyses of cell polarity can give hints about the nature of the gradient. For example, if squares of *Rhodnius* are rotated and the results compared with computer simulations of concentration gradients that are allowed to diffuse with time, the fit is only reasonably good. Also, there is independent evidence that the pattern does not change through the passage of time alone. If adult *Rhodnius* are made to moult again after several months, by injecting them with hormones, they make a new cuticle with an unchanged pattern — suggesting the polarities of the cells may be at some equilibrium. If the cells divide between the two adult moults the patterns do change, suggesting that the equilibrium can change at cell divisions. Computer models of equilibrium gradients also give a better fit to observed patterns.

Figure 6.7 shows a computer model which is applied to 90° and 180° rotations, to be compared with photographs of *Rhodnius* adult cuticle after the operations. For the model, a simple concentration gradient is assumed and the cells are given a particular strength in their ability to maintain their prior concentration. This parameter has two extremes; at one they succeed in maintaining their original concentration com-

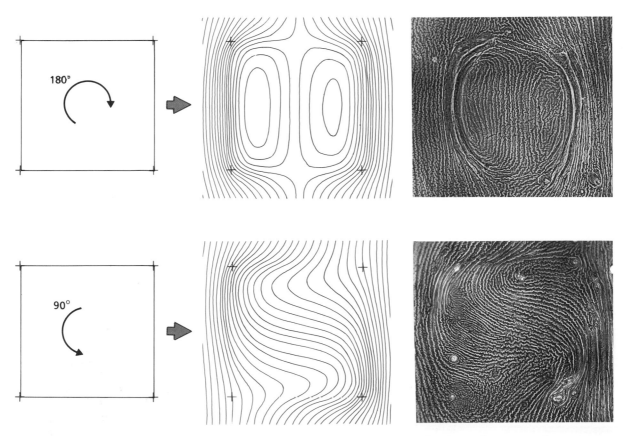

Figure 6.7 Modelling diffusion gradients. Predicted and actual outcomes from rotation experiments in *Rhodnius*.

pletely after the operations and this would give a precipitous landscape, reminiscent of Manhattan. At the other extreme, diffusion will lead to the complete decay of the perturbation produced by the operation. Intermediate values give intermediate patterns at equilibrium. The chosen value (and it is the only important parameter the modeller can vary at will, the others being set by the diffusion equations) gives the patterns shown for 90° and 180° rotations. Of course this does not prove that the gradient is a diffusing morphogen, but it shows, at least, that the model is adequate to give the patterns observed. There are also other minimal conclusions that can be drawn: polarity is not sufficient to explain all the results; there is some quantitative difference between cells at different points in the anteroposterior axis of the segment — they 'know' where they are. Positional information is not immutably fixed in a cell, it can be changed by cells nearby, the cell taking up a numerical average of the cells in its neighbourhood; it behaves as a continuous variable. Models which resort to a series of discontinuous and unrelated labels would have to be made fantastically ornate to mimic the experimental results.

This gradient model provides a partial explanation for the detailed patterning of groups of cells and for their polarity. In the anteroposterior axis there is considerable precision — witness the narrow ridge in *Galleria* (perhaps this is why most patterning in the abdomen of insects is stripy, with the stripes oriented in the mediolateral axis, as in a wasp). One can see this too in the cuticle of the *Drosophila* abdomen, where the predominant dorsal pattern is a series of zones arranged along the anteroposterior axis. However, insects are more complex than that; there has to be specification in the mediolateral axis, as well. Along the dorsal midline of many insects, there is a very narrow line of cuticle that is specifically weakened, so that it will split at moulting. Also, the well named bug, *Graphosoma*, illustrates that stripes can be made parallel to the anteroposterior axis (Figure 6.8). So

Figure 6.8 *Graphosoma italicum*, a bug with a checkerboard of anteroposterior and mediolateral stripes.

far, there are no models to explain this mediolateral patterning; perhaps, as in the egg, specification in the two main axes of the body will prove to be independent and distinct.

Gradient, segments and parasegments

There is a relationship between compartments and gradient fields. Studies of regeneration on *Oncopeltus*, the milkweed bug, show that as cells grow to replace pieces of the segment or to intercalate intermediate gradient values, their cell lineage is constrained. For example, if the segment border is being restored following ablation, then it always forms at the interface where cells of the two segments meet as they migrate towards each other. Figure 6.9 shows how the experiment is done. In the first step a piece of epidermis is transplanted to create a marked patch. In the second, the graft boundary and the segment boundary are both cauterised, so that cells on either side will migrate in to heal the wound. At the anterior edge of the graft an uneven boundary forms, but at the posterior edge a segment boundary is reconstructed. This consists of a straight-edged interface between the cells of different segmental origin, with not a single cell straying across the boundary. Of course, regeneration is different from normal development, but these experiments reveal a relationship between lineage and gradient boundaries that probably exists in the embryo too. In my view, the lineage boundaries established by *ftz*, *eve* and *engrailed* in the *Drosophila* embryo are likely to register gradients of positional information, but the form of the gradient landscapes and their relation to segments and parasegments is unknown. There are a number of possibilities, but the one I prefer is a sawtooth gradient landscape with alternating direction of slopes in the anterior and posterior compartments; changes of slope occurring at the boundaries (see Figure 6.3). This is compatible with the experiments on bugs because, in those where the posterior compartment has been studied, it is very small and would have gone unnoticed in most of the grafting experiments. Figure

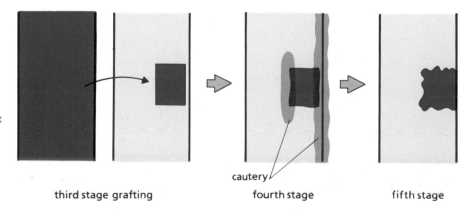

Figure 6.9 In *Oncopeltus*, cell originating in different parts of adjacent segments meet and form a straight interface. Cells from nearby parts of the same segment meet at a wiggly interface.

third stage grafting   cautery   fourth stage   fifth stage

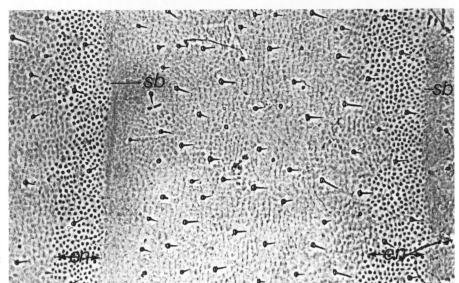

Figure 6.10 The segment of *Oncopeltus*. A narrow strip of posterior nuclei stain for the *engrailed* (*en*) antigen; the posterior limit of that strip defines the segment boundary (sb).

6.10 shows the posterior compartment of *Oncopeltus*, as revealed by *engrailed* expression.

There will probably be gene products required to read polarity and to express it. A candidate is an integral membrane protein encoded by the *frizzled* gene. In *frizzled⁻* flies the orientation of bristles and denticles is awry — when small patches of *frizzled⁻* clones are made in the wing, the denticles form wrongly oriented whorls; distal to the clones the wildtype cells become disoriented. It seems that the *frizzled* gene is required for the passage of some polarising signal from proximal to distal along the wing — and elsewhere.

I have gone into positional information in detail because, although currently out of fashion, it is important and will probably apply to *Drosophila* and elsewhere. Some or many of the segment polarity genes (p. 101) will probably be involved in establishing and maintaining systems of positional information — as their mutant phenotypes, with their bewildering changes in polarity and pattern, suggest. The molecular study of these genes will probably be the means to bring gradient theories, at present irritatingly abstract, to a more molecular and concrete state.

Competition and compartments

Analysis of clones in *Drosophila* has given an important insight into growth that deserves to be better known. It began with the use of *Minute⁺* clones, when it was noticed that, not only did the clone grow faster than the *Minute⁺/Minute⁻* cells, but that the clone could almost fill the compartment — implying the **elimination** of the *Minute⁻* cells by competition between stronger and weaker cells. This finding was followed up: it was found that cells within the same compartment do compete with each other. Two *Minutes* are used — *Minute^A* and *Minute^B*, that are on different chromosomes — both produce similar developmental delay and flies reach maturity about 2 days later than wildtype sibling flies in the same bottles. If *Minute^A*/+ clones are produced in a background of wildtype cells they are eliminated by competition and are not seen in the adult. If the same *Minute^A*/+ clones are produced in sister flies that are mutant for *Minute^B* they develop well — for they are now growing up amongst cells that are disadvantaged to a similar extent as they.

The mechanism of cell competition is unknown but it is clear it is a local phenomenon and this is shown by some ingenious experiments. The parent flies are constructed so as to generate embryos which will have two types of clone that are differently marked and induced at different times; one will be *Minute$^+$* and grow large in the *Minute$^+$*/ *Minute$^-$* background (pink, Figure 6.11), the other will be simply a marked clone that does not change the *Minute* properties of its background. These latter clones are studied: if they are subclones within the *Minute$^+$* clone, but near its edge (1, Figure 6.11) they tend to be particularly large, implying rapid growth at the edge of the clone. Subclones fully inside (2) are smaller but still larger than those *Minute$^+$*/ *Minute$^-$* clones outside the *Minute$^+$* clone (3,4). Clones just outside the invading edge of the *Minute$^+$* clone are small and broken up into pieces, as if they are being destroyed (5), while any clones across the compartment border from the *Minute$^+$* clone are normal (6).

The picture is of some overall control of the number of cells that is achieved both by influencing the amount of cell division and by cell death. The number of cell divisions is not counted, thus clones developed from sister cells in the wing usually have different numbers of cells in them, and if cells are eliminated by random cell death (e.g. X-rays) extra divisions compensate for the loss. The cells that do divide and/or the cells that die are chosen by some competitive process, such that the weaker ones in the group are eliminated while the stronger ones divide.

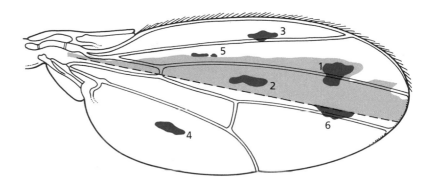

Figure 6.11 Local effects in competitive growth. A large *Minute$^+$*/*Minute$^+$* clone has grown in the wing (pink). The behaviour of smaller, later induced clones can then be studied.

This complex process achieves considerable precision. For example, left and right wings, which grow independently from the blastoderm stage, are very similar in size and shape. Also, in all insects, the net increase in dimensions of parts of the body is very precisely regulated. If the increase in length of a leg segment of a cockroach is plotted on a log scale against the number of larval stages, an extraordinarily straight line is the outcome. Each part grows at its own characteristic rate; when different parts grow at the same rate proportional growth is the result, when they grow at different rates disproportionate or allometric growth follows (see Figures 6.13–6.15). Studies of growth of this type give the impression of some mathematically precise control which operates independently in different body parts.

There are two aspects of growth control I want to consider further: the relationship with gradients and the genetics of shape, size and proportion.

Growth and gradients

Again we have to turn away from *Drosophila* to other insects, this time to a cockroach. In a long series of experiments it was shown that, if pieces are cut out from the middle of the tibia, over a moult or two the missing pattern is regenerated to the correct length. This is apparently done with reference to a segmental gradient which is present in the limb and is similar to the gradients described above in the abdomen. Cutting pieces out from the tibia puts cells from different levels in apposition, growth is stimulated and excessive cell division continues locally until length is restored (Figure 6.12A). These extra cell divisions occur near to the junction between the host and the graft and generate a section of tissue (red) to replace the gap — this is known as intercalary regeneration. Individual segments that are extra long can be made, experimentally, by confronting distal and proximal parts of the tibia as shown in Figure 6.12B, and the resulting intercalary regenerate has, as would be expected, an opposite polarity. The implication of these experiments is that somehow information as to tibial length is conveyed by the gradient; the second experiment suggesting that the length is not measured as such but that some correlate of length is perceived and acted on locally. There is, in theory, a simple way to do this: if the limits of the gradient, the boundary conditions, are fixed and the slope is a monotonic decline between those limits, then the slope provides a measure of the length of the whole and, in principle, this can be perceived **locally**. Thus when slopes are steepened by experiment, cell division rates should locally increase, as observed. Note that this machine depends on cell interaction leading the process: placing level 7 next to level 5 results in some blending, a sudden decline from 7 through 6 to 5 that precipitates the extra cell division.

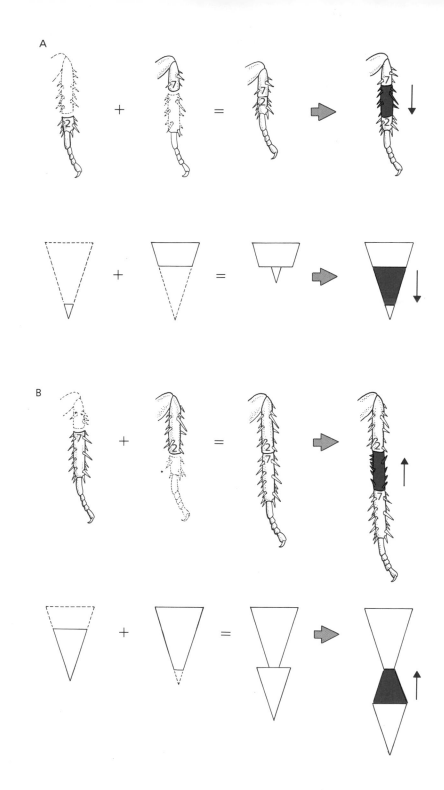

Figure 6.12 Experiments on cockroach legs. Intercalary regenerates are shown in red with both the results and the gradient interpretation given.

The reverse situation may also occur: grafting experiments or mutations may set up unusually flat gradients. Then the number of cell divisions might well be reduced or eliminated and the amount of cell death increased; the result of both being a shift towards a shorter structure and a steeper gradient. This type of size regulation may be the cause of cell death in mutants with missing gap genes or pair rule genes. For example, in *ftz⁻* embryos, extra large parasegments are formed because alternate boundaries are missing. The result is a zone of excessive cell death in each giant parasegment and, eventually, a small embryo.

Taken together, all the experiments on insect segmental gradients suggest cells can respond to the scalar level of the gradient (position), the vector in the gradient (polarity) and the first derivative or steepness of the gradient (growth).

Shape and genetics

The determination of shape and proportion has been little explored. There must be genetic control of shape, although it is not known if there are 'shape genes' or whether shape is the indirect outcome of many processes. When species evolve, the critical changes are not alterations of occasional amino acids in major proteins such as structural proteins or enzymes, but changes in form. For example, when humans and chimpanzees are compared, their proteins are found to be more than 99% identical in amino acid sequence. Since many of even these changes have no functional consequence it is hard to credit that they could be responsible for the morphological differences between the two species. Some groups of species are very conservative and change little anatomically over long geological periods, others evolve at high rates. Yet when rates of change of protein sequences are compared in slow and fast evolving groups they are found to be about the same. Therefore evolution must work on other factors that determine shape and proportion, but what are they? There is no doubt that shape and pattern respond very rapidly to selection — dogs are an impressive example. Sometimes, shape changes are so consistent that they give the impression that they have an inertia, are propelled along a path. The impression may be misleading of course, but look at Figure 6.13.

The way the shape and pattern of insects evolves can tell us something about how shapes are moulded during development. For example, if one pair of legs can enlarge while the other pairs remain unchanged, the different segments must be under independent genetic control — for evolution must work by manipulating genes or the controlling regions of genes. If a pair of legs can fatten but not lengthen we learn that the different leg axes can be controlled independently, too. Look at Figure 6.14, the legs of two related species of Ephydrid flies are shown, one is predatory, one is not. On the left, note that the prehensile legs of T1 of *Octhera mantis* are specialised while the T2 and T3 legs strongly resemble the legs of the peaceable *Parydra aquila*.

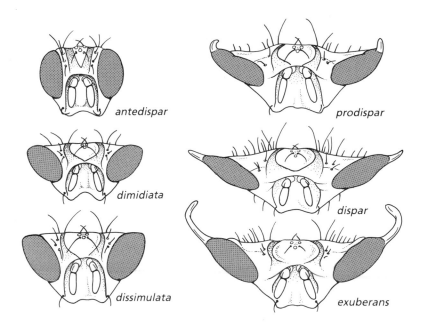

Figure 6.13 Heads of related species of *Drosophila*. These can be arranged in a series stretching from the mundane to the fantastical.

antedispar

prodispar

dimidiata

dispar

dissimulata

exuberans

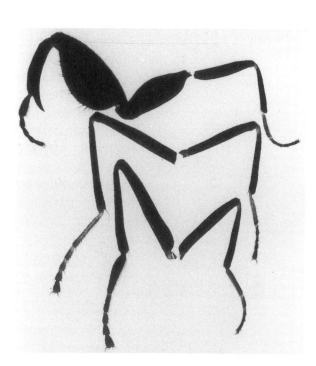

Figure 6.14 The legs of two related flies. On the left a predatory species and, on the right, the equivalent legs of a peaceful relative.

One does not need to marvel at the diversity of shapes of insects for long to conclude that genetic mechanisms have placed few constraints on evolutionary radiation. One constraint is apparent, however — insects are remarkably bilaterally symmetric, there appear to be no 'left' or 'right' genes. An example of bilateral asymmetry is the genitalia of the male; as they develop they rotate — always counterclockwise 180°. It

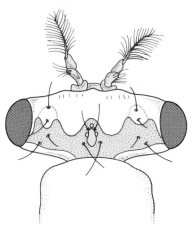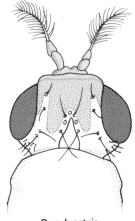

Figure 6.15 Two species of *Drosophila* that, in spite of their strangely shaped heads, will interbreed.

D. heteroneura　　　　　*D. sylvestris*

is not known what gives direction to the rotation. Another is that if many different *Drosophila* species are compared it is found that while they vary considerably in size (a five-fold difference in wing area has been recorded), the proportion and shape of the parts is much more conserved. This might suggest that genetic mechanisms to scale up or down are more available than those that deform, but shape changes can also be rapid: there are two species of *Drosophila*, *D. heteroneura* and *D. silvestris*, that live on Hawaii. They are very closely related; for example if amino acid sequence similarity of 12 enzymes is estimated by electrophoresis, they appear to be among the most related pairs of species on the island. In spite of this, they look very different, the head of the male *heteroneura* looking like that of a hammer-head shark (Figure 6.15). Although they rarely interbreed in the wild, they will do so in the lab, giving fertile offspring. From a quantitative analysis of the progeny it appears that rather a small number of genes is responsible for the difference in head shape, probably about 10. It would be extremely interesting to know what these genes are: are they alleles of genes that are already known from more drastic mutant phenotypes, or do they belong to a yet undescribed class of shape genes?

Another intriguing aspect of evolutionary changes is that while some organs remain untouched, others alter with great freedom. Taxonomists use genitalia to classify insects such as beetles because they seem to be made of a malleable material that changes shape with alacrity — by contrast, the shapes of the guts of a family of beetles are highly conserved. Of course, it is not clear whether stagnation is due to lack of ability or lack of need; probably it is a bit of both, since increased sophistication of organs, such as the central nervous system, will probably go hand in hand with increased flexibility in evolution. Rapid evolutionary changes of shape and proportion take us back to the

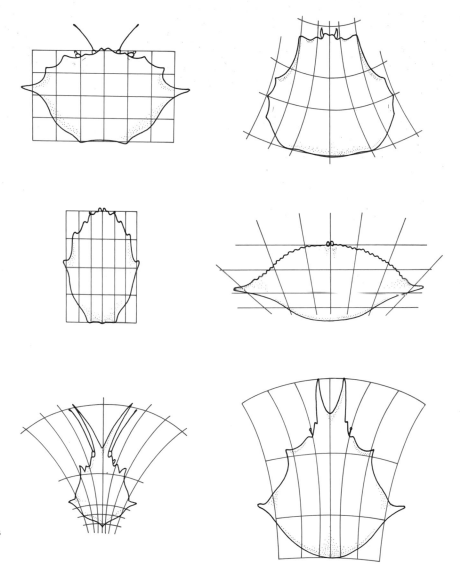

Figure 6.16 Thompson's crabs arranged on a common coordinate grid.

observations of Thompson, who was fond of comparing the forms of related animals and showing how an organ of one species could be transformed into that of another by simple proportionate changes in the different axes (Figure 6.16). Again, it is not possible to be sure whether these comparisons of crabs and skulls, etc. are facile or fundamental. Thompson thought the latter: 'There is something ... indispensable ... which is common to them all. In these transformations ... every point and every line ... keeps its **relative** order and position throughout all distortions ...' [my emphasis]. Thus he thought 'variation has proceeded on definite and orderly lines, that a comprehensive "law of growth" has pervaded the whole structure.' [10]

Perhaps a few shape genes that work at the level of **gradient** interpretation are responsible for these 'laws of growth' and it is these that would be the immediate targets of any evolutionary pressure for morphological change. It is obvious that we barely know enough about the determinants of shape during development of one species to speculate about changes of shape between species. It is just that thinking about the latter may help direct experiments on the former.

Further reading

REVIEWS

King, M.C. and Wilson, A.C. (1975) Evolution at two levels in humans and chimpanzees. *Science* **188**: 107–116.

Lawrence, P.A. (1973) The development of spatial patterns in the integument of insects. In: Counce, S.J. and Waddington, C.H. (eds) *Developmental Systems: Insects*. Vol. 2, pp. 157–209. Academic Press, London.

Thompson, D.W. (1942) *Growth and Form*. 2nd edn. Cambridge University Press, Cambridge.

SELECTED PAPERS

Competition

Simpson, P. and Morata, G. (1981) Differential mitotic rates and patterns of growth in compartments in the *Drosophila* wing. *Dev. Biol.* **85**: 299–308.

Evolution of shape

Grimaldi, D. (1987) Phylogenetics and taxonomy of *Zygothrica* (Diptera: Drosophilidae). *Bull. Am. Nat. Hist.* **186**: 103–268.

Val, F.C. (1977) Genetic analysis of the morphological differences between two interfertile species of Hawaiian *Drosophila*. *Evolution* **31**: 611–629.

frizzled

Vinson, C.R., Conover, S. and Adler, P.N. (1989) A *Drosophila* tissue polarity locus encodes a protein containing seven potential transmembrane domains. *Nature* **338**: 263–264.

Gradients and polarity

Campbell, G.L. and Caveney, S. (1989) *engrailed* gene expression in the abdominal segment of *Oncopeltus*: gradients and cell states in the insect segment. *Development* **106**: 727–737.

Lawrence, P.A., Crick, F.H.C. and Munro, M. (1972) A gradient of positional information in an insect, *Rhodnius*. *J. Cell Sci.* **11**: 815–853.

Locke, M. (1959) The cuticular pattern in an insect, *Rhodnius prolixus* Stål. *J. Exp. Biol.* **36**: 459–477.

Marcus, W. (1962) Untersuchungen über die Polarität der Rumpfhaut von Schmetterlingen. *Wilhelm Roux's Archives* **154**: 56–102.

Nübler-Jung, K. (1979) Pattern stability in the insect segment. II. The intersegmental region. *Wilhelm Roux's Archives* **186**: 211–233.

Stumpf, H.F. (1968) Further studies on gradient-dependent diversification in the pupal cuticle of *Galleria mellonella*. *J. Exp. Biol.* **49**: 49–60.

Wright, D.A. and Lawrence, P.A. (1981) Regeneration of the segment boundary in *Oncopeltus*. *Dev. Biol.* **85**: 317–327.

Growth

Bohn, H. (1974) Extent and properties of the regeneration field in the larval legs of cockroaches (*Leucophaea maderae*). III. Origin of the tissues and determination of symmetry properties in the regenerates. *J. Embryol. Exp. Morph.* **32**: 81–98.

SOURCES OF FIGURES

For details, see above.

Figure 6.2 See Lawrence (1973) and Locke (1959).

Figure 6.3 See Lawrence (1973).

Figure 6.4 After Stumpf (1968).

Figure 6.5 See Nübler-Jung (1979); drawing based on my own unpublished grafts.

Figure 6.6 After Lawrence (1973) and Marcus (1962).

Figure 6.7 After Lawrence *et al.* (1972).

Figure 6.8 Drawing by Denys Ovenden.

Figure 6.9 After Wright and Lawrence (1981).

Figure 6.10 From Campbell and Caveney (1989).

Figure 6.11 After Simpson and Morata (1981).

Figure 6.12 After Bohn (1974).

Figure 6.13 After Grimaldi (1987).

Figure 6.14 Insects from Henry Disney.

Figure 6.15 After Val (1977).

Figure 6.16 After Thompson (1942).

7 Spacing patterns

EMBRYONIC CELLS of the central and peripheral nervous system as well as precursors of the sensory organs of the adult are selected from the epithelium on the basis of their position. First, a group of cells is identified, in part by the localised expression of one or more transcripts of the achaete-scute complex. Second, competitive interactions between the group select one cell of their number, a process directly involving the Notch gene.

Spacing patterns and neurogenesis

Up to now we have considered the ground plan of the embryo and how it is laid down in scale and proportion, but have not thought much about how detailed patterns are formed. Above all things, insects are **intricate**; consider a Mymarid wasp — a minute insect, less than 1 mm long, that completes its entire life cycle in the egg of another insect — which is able to fly, to find a mate, to locate and identify the eggs of its host and has all the necessary equipment and control systems packed into its tiny body. One ultimate aim of the embryologist is to understand how precision organisms like this can be built. The way things are made may be much less complex than the final outcome; underlying the bewildering variety of pattern and form in biology (consider butterfly wings!) there are **relatively** simple mechanisms that are used again and again. In butterflies there is even a reason for believing this, for changes in only one gene can produce several patterns which, to our eyes, look completely different (Plate 7.1).

One of the mechanisms that is used repeatedly is the selection of a subset of cells from a larger homogeneous group, so that two cell types are produced from one, and in a particular pattern. The simplest of these is exemplified by bristles in the insect epidermis. In the most common form the bristles derive from a minority of cells that are **overdispersed**, that is arranged in a pattern that maximises the distance between them. This is called a spacing pattern and turns up again and again in nature. There is both an experimental (other insects) and a genetic (*Drosophila*) approach to understanding how this pattern is made. In *Oncopeltus*, during the metamorphosis from larva to adult, an even sheet of epidermal cells with very few bristles becomes a dense mat of bristles that are evenly spaced. This process has two separate aspects; first, the selection of an overdispersed set of epidermal cells which become transformed into bristle mother cells; and second, the division of the mother cells to give the bristle components. We return to the second aspect on p. 172.

Spacing patterns — a model

Consider the first problem: what is observed in various insects is that the bristle mother cells always appear approximately in the centre of the spaces between the extant bristles and this occurs continuously until the final density is reached. So whatever mechanism is responsible is a positional one; cells are not chosen at random, but because they are far enough from existing bristles. Wigglesworth carefully drew the bristles on part of one segment of a *Rhodnius* larva at the fourth stage and compared the same area following moulting of the same individual to the fifth stage (Figure 7.1). It was clear that new bristles had been added and that these had appeared in the largest spaces between the preexisting bristles. Even within one moult cycle, bristles are added in a non-random order: studies on *Oncopeltus*, where the bristle density increases about 10-fold during metamorphosis to the adult, show that the space is occupied in order, the new bristles appearing first in approximately the centre of the largest spaces. Four stages in the process are shown in Figure 7.2.

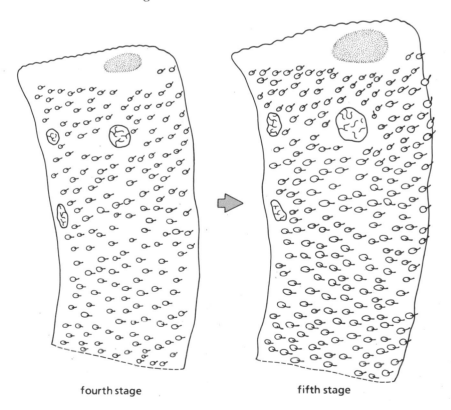

Figure 7.1 In *Rhodnius* new bristles (in red) are added in the spaces between extant ones.

fourth stage

fifth stage

Experiments in which cell density is varied show that the distance is not measured absolutely, but in numbers of cells. The simplest kind of model was proposed by Wigglesworth; he suggested that there might be a substance produced by the bristles that diffuses and inhibits other epidermal cells from becoming bristles. Once the concentration of this substance falls below some threshold, then an epidermal cell will transform into a bristle mother cell, start to develop as a bristle, begin to produce the substance, and inhibit other nearby cells from becoming bristles. In normal growth, the epidermal cells increase in number and the concentration landscape changes so that, furthest from existing bristles, the concentration of inhibitor will be lowest and, there, the threshold will be reached first.

There is a fundamental difficulty with this model, which is that both common sense and computer simulations suggest that frequently several cells would fall below the threshold simultaneously; this would give clusters of adjacent bristles, and these are not observed. Some element of dynamic **competition** is needed to make sure that only one of these presumptive bristles wins through, and this has been introduced into the model as follows: consider an inhibitory substance that diffuses in the epithelial sheet and stops cells from becoming bristles; bristles are net producers and epidermal cells net absorbers of the substance. It is assumed that an uninhibited cell has an intrinsic propensity to become a bristle, but the more bristle-like it becomes, the more inhibitor

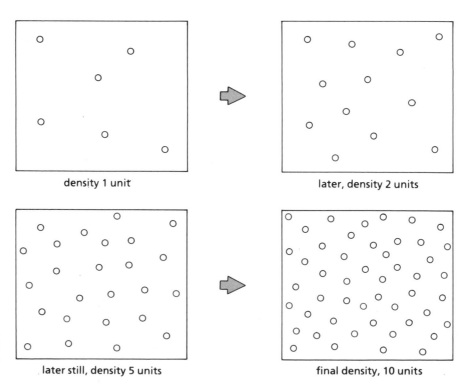

density 1 unit

later, density 2 units

later still, density 5 units

final density, 10 units

Figure 7.2 In *Oncopeltus*, bristles are added in a non-random order, the centre of the largest available spaces always being preferred.

it produces and, if it is to go on, it must have nearby epidermal cells to absorb the inhibitor. I should emphasise the words 'if it is to go on', which implies that epidermal cells do not become bristle mother cells in an instant; rather, there is a progressive advance towards full commitment. During this advance, if circumstances change, the trend can be reversed and the cell returns to being an epidermal cell.

This model was invented to explain a one-dimensional spacing pattern in a blue-green alga, where the gradual acquisition to commitment was nicely documented. The model is illustrated diagrammatically in Figure 7.3. The alga consists of a filament with dispersed special H cells. The H cells do not divide but the other cells do and the filament lengthens; as it does so occasional cells convert and form new H cells. In the model there is an inhibitor Y and this diffuses along the filament; it is produced mainly by the H cells and forms gradients declining away from each. As the filaments lengthen, a number of cells are released from inhibition and begin to develop towards the II state. Some turn grey in colour, making it obvious that several cells are involved; only one of these can win through and the others will regress as they are inhibited by the winner. Observation and experiments support the model; even in normal development, cells are observed to begin differentiating from their neighbours and then reverse. Also, if considerably differentiated and apparently committed cells are experimentally separated from the algal filament, with only one or two cells to provide 'support' (in terms of the model, to absorb inhibitor) they then revert to the ordinary cell state. Cells which are too far differentiated

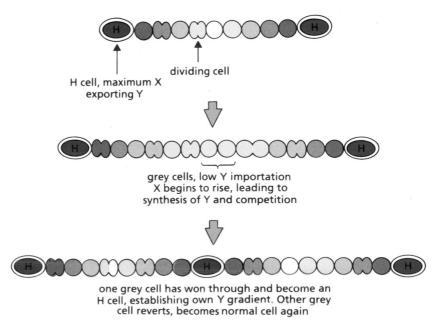

Figure 7.3 *Anabaena*, a one-dimensional pattern. As cells divide, new H cells arise, far from preexisting H cells. The gradient of inhibitor Y emanating from H cells is indicated by the red colour. H cells are imagined to be maximally differentiated (have maximum 'X'); the production of Y correlating with the amount of X.

H cell, maximum X exporting Y

dividing cell

grey cells, low Y importation X begins to rise, leading to synthesis of Y and competition

one grey cell has won through and become an H cell, establishing own Y gradient. Other grey cell reverts, becomes normal cell again to revert in this test can still change back if they are completely isolated, with no support.

These experiments on the blue-green alga allow two important generalisations about development which should be borne in mind when reading the literature: first, the acquisition of a new cell state, to become 'determined', is probably a gradual process. Second, attempts to find out when cells are determined are likely to yield variable answers, for a cell that is intransigently committed by one experimental test may be pliable by another — an inconvenient fact of life that has led to numerous disputes amongst embryologists. The model for the blue-green alga can also be applied to the distribution in two dimensions of insect bristles or even mammalian hair follicles. It achieves what is generally referred to as **lateral inhibition**, the process where one developing element prohibits the development of similar elements nearby.

The model needs to be made more concrete, but one can see that it could work in principle. A key element is the diffusing signal; this need not be a small molecule that can penetrate cell membranes, it could be a secreted protein that would bind to a receptor on the responding cells. More has been learnt about lateral inhibition from more defined patterns of bristles in the *Drosophila* adult and, also, the defined pattern of neuroblasts in the embryo.

Landmark bristles

Some *Drosophila* bristles do conform with a simple spacing pattern, such as the smaller ones on the thorax, legs and abdomen, but others are found in such constant positions in different flies they are like landmarks and have been given names. For example, there are two bristles on each side of the thorax that are called the anterior and posterior dorsocentrals and every wildtype fly has them. Bristles like these must be produced by a more sophisticated system and the type of model currently favoured is a two-step one. The idea is that in the first step, a group of cells, a **proneural cluster**, is allocated in more or less the right place and then, in the second step, the cells of the cluster compete with each other to select one of their number in a similar procedure to that in a small region of the spacing pattern (e.g. amongst the 'grey' cells in Figure 7.3). We consider these steps in turn: first, how is the proneural cluster positioned and the cells allocated to it? Later, we ask how one cell from the cluster is selected.

The achaete-scute complex is instrumental in the positioning of proneural clusters. This gene complex (AS-C) has extraordinary mutant phenotypes. Complete lack of the AS-C eliminates most, but not all, of the central nervous system in the embryo and in clones eliminates most, but not all, adult bristles. Partial mutant alleles remove particular sets of bristles in a confusing mixture of overlapping sets; there is no simple hierarchy. I mean that if a weak mutant removes only bristle A, a stronger one may remove B, C and D, but leave A intact.

Molecular analysis of the AS-C shows that it has a long and complex regulatory region and encodes as many as four closely related proteins with overlapping functions. The close relationship of these proteins is discovered when the amino acid sequences are compared; in a particular region there are common parts which are the signature of a 'helix–loop–helix' element. This is a structural element in the protein that is thought to allow it to recognise and bind to specific sequences in the DNA. *In vitro*, helix–loop–helix proteins have the special property that they only bind to DNA when they are associated in pairs, often as heterodimers made up of two different members of the family. There are other such proteins in the genome, and one of these is the product of the *daughterless* gene; since *daughterless⁻* embryos have no peripheral nervous system, it is thought that the *daughterless* protein might form heterodimers with any of the AS-C proteins. The *daughterless* protein is ubiquitous, so effective heterodimers will be dependent on, and located solely by, the patterned expression of the AS-C genes. Accordingly, the AS-C gene products are expressed in precise and partially overlapping patterns, both in the embryo, where neurogenesis occurs, and in the imaginal discs when bristles are forming. The distribution of the RNA suggests that the genes are expressed in the cells of proneural clusters. This has been demonstrated in the wing imaginal disc where the pattern of RNA fits expectations. First, some few bristle sites express RNA, but later more and more are seen until, finally, the numerous proneural clusters for the overdispersed bristles appear. The process extends over about 2 days.

The cells constructing these bristles can be highlighted by employing a 'blue jump' stock in which β-galactosidase is activated only in bristle mother cells (see Box 3.3, p. 62). The *scute* RNA in the same disc can be shown up by *in situ* hybridisation. What is found is that, in some of the places where bristles will develop, there is a cluster of cells expressing the *scute* RNA. A few hours later, one of these cells enlarges and begins to express β-galactosidase. Plate 7.2 shows two discs that are from mature larvae; (a) is a few hours younger than (b). The chocolate colour marks *scute* RNA and the blue colour the β-galactosidase. If you compare the two pictures you can see how areas expressing *scute* RNA forecast the appearance of one or more bristle mother cells. Some bristle mother cells have been formed even earlier and can be seen already in the younger disc.

Studies with the reporter gene *lacZ* have shown that *scute* expression in the cells of the proneural cluster and in the bristle precursor itself are controlled by separable regulatory elements. Once the specific cell has been chosen, it continues to express *scute*; consequently the concentration of gene product in it may reach a higher level than in the remaining cells of the cluster.

It will be no easy matter to discover how the different transcripts of the AS-C are regulated and positioned correctly; nor why there are several of them. It will also be necessary to understand the role of the other homologous genes. Certainly, the four transcripts of the AS-C do have different roles, as mutants are known that eliminate them selectively. For example, *scute*⁻ and *achaete*⁻ are classes of mutations which each eliminate one transcript. Amongst other tasks, these transcripts are needed for the formation of bristles on the adult thorax: *achaete*⁻ mutations remove three of the landmark bristles and all the overdispersed smaller ones, whereas *scute*⁻ mutations eliminate nine of the landmark bristles and leave the smaller bristles untouched. Weaker *scute* mutations eliminate only some of the bristles but those that remain are positioned as normal — suggesting that whatever places the *scute* transcripts does it bristle by bristle: indeed in those weaker *scute* mutations some of the patches of the transcript called T4 are omitted, and these correspond with the missing bristles.

It is important to note that the selection of bristles occurs serially and in a background of dividing cells. Pieces of evidence from different systems suggest that the larger sparser bristles are determined first and as cell divisions of the intervening cells continue they make room for a second round of more numerous proneural clusters. Accordingly, patches of cells expressing AS-C do so at different times in the imaginal disc and, in the embryo, the pattern of *scute* transcripts is continuously changing.

It would be simplest if *scute* expression were sufficient to position bristle development; however this is not so. If *scute* transcripts are expressed universally in transformed flies carrying a *scute* gene under the control of a heat shock promoter, bristles do not form everywhere. Even more persuasively, ubiquitous *scute* expression in flies that lack both the *achaete* and *scute* genes leads to the formation of some normally placed bristles. Thus *scute* expression is necessary for bristle formation, but other factors are also involved in timing and positioning.

As might be expected, there are several other genes whose products are needed for the sophisticated process of regulating AS-C transcription. One example is *hairy*; this mutation causes *scute* transcripts to be expressed in the wrong places, turning a bald area into a bristly one. The bristles are not chaotically spaced; they are still overdispersed, suggesting that the competitive process is working normally.

Competition — the evidence

We now turn again to the question, how is it that only one of the candidate cells in each cluster is chosen? How is it that, in an even field of bristles, the bristles are spaced apart from one another? The evidence for a competitive selection of bristle cells and neurones has come from several sources, both experimental and genetic. The most telling experiment was carried out on the development of a particular neuroblast in the locust embryo. If all the cluster of nearby cells as well as the presumptive neuroblast are killed with a laser beam, the neuroblast does not form, showing that it cannot be replaced indiscriminately. However, if only the neuroblast cell is killed just as it begins to grow, another neighbouring cell almost always takes its place and produces the neuroblast. This shows that the potential is present in several cells and that, normally, only one of those wins through to make the neuroblast, inhibiting the others from doing the same.

More evidence for competition between candidate cells comes from classic genetic experiments by Stern — he used gynandromorphs of *Drosophila*, genetic mosaics in which the male tissue was *achaete*⁻ and the female tissue *achaete*⁺. The male cells were marked with the *yellow* mutation, the female cells carried *yellow*⁺. He studied particular landmark bristles that are always present in *achaete*⁺ flies. Whenever the female tissue included the site of a bristle, the bristle always formed, even if there was only a little isolated patch of *achaete*⁺ cells, or if the bristle site was on the extreme edge of the female tissue. Four examples are shown in Figure 7.4, where the pink cuticle and red bristles are female, *yellow*⁺ and *achaete*⁺. The male, *yellow*⁻, *achaete*⁻ tissue is shown in grey and empty bristle sites as open circles. The result shown in Figure 7.4A tells us that the *achaete*⁻ mutant removes the bristle, not by changing the overall organisation of the disc, but by specifically removing the propensity to form some bristles but not others.

Stern also found that when the bristle site was male and *achaete*⁻, the bristle was missing; whatever the mosaic pattern, no male cell ever made the landmark bristle (Figure 7.4). However, when there were female *achaete*⁺ cells near to the site they occasionally made a single, slightly mislocated bristle (which was, of course, *yellow*⁺ in genotype) (Figure 7.4B). Stern's interpretation was that the loss of the landmark bristle eliminated the lateral inhibition that it would normally produce, allowing any nearby cell that was genetically competent to produce a substitute bristle. Historically, this was the first evidence that there are several cells (now called a proneural cluster) that are each capable of producing a single landmark bristle.

To extrapolate from Stern's experiments and find out more about lateral inhibition we must introduce some new genes. Note that many of the genes have a role both in the selection of the bristles (being part of the peripheral nervous system) and the neuroblasts (the central nervous system). The reason must be that these two processes have much in common, both have to select a certain number of cells from

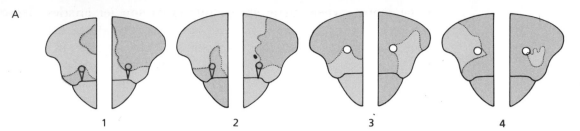

A

1 and 2: whenever the site is female, *achaete*⁺ and *yellow*⁺ the bristle develops

3 and 4: usually, whenever the site is male, *achaete*[−] and *yellow*[−], the bristle is absent

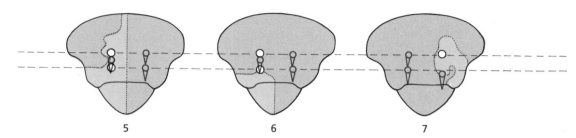

B

three thoraces in which on the mosaic side there is a single displaced bristle, in the *achaete*⁺ territory,
that has formed instead of the normal bristles

Figure 7.4 Stern's gynandromorphs. The bristles studied are the anterior and posterior dorsocentrals; these are always present in their correct sites in the wildtype and missing in *achaete*[−] flies. Parts 1 and 2 feature the posterior dorsocentrals, which are present in the four half-thoraces shown. The four half-thoraces in 3 and 4 have missing anterior dorsocentrals. Parts 5−7 show three complete thoraces, in each one side only is affected; there only a single bristle forms. Female, *achaete*⁺ tissue is shown in red, male *achaete*[−] tissue in grey.

an epithelial sheet on the basis of position, and both have to pattern identifiable nerve cells or bristle cells (compare also the selection of ommatidial clusters in the spacing pattern of the eye, p. 186). There are several genes which are needed for the generation of two cell types, so that in the complete absence of their products, all or most of the competent cells form nerve cells in the embryo or bristles in the adult. Examples are the *Notch*, *Delta* and *shaggy* genes. Mutants of these three genes are embryonic lethals and, in the absence of maternally derived gene product, give their strongest phenotype. In *Notch*[−] and *Delta*[−], that part of the ectoderm that would normally contribute to the central nervous system makes a mass of neuroblasts and no epidermal cells, while the dorsal ectoderm (which does not make neuroblasts) is little affected.

In the case of *shaggy*[−], the embryos try to develop with **all** the cells forming neuroblasts. *shaggy*[−] cells are viable in patches in the adult and make supernumerary bristles. The pattern is specific; extra bristles are added only where bristles are normally found, for example a landmark bristle may be replaced by a group of similar bristles. Elsewhere in the thorax, where there are overdispersed smaller bristles, dense masses of extra bristles are found, but, in areas where the cuticle is free of bristles, normal cuticle is made. In flies carrying weak *scute* mutant alleles that eliminate particular bristles, *shaggy*⁻ clones do not produce clusters of bristles in place of those particular ones. The simplest explanation for the mutant phenotype is that, in *shaggy*⁻ clones, lateral inhibition fails and many of the cells of the proneural clusters proceed to make bristles.

A 'similar interpretation applies to *Notch*⁻ mutants. The ventral ectoderm of the embryo generates both neuroblasts and epidermis, while the dorsal epidermis makes only epidermis and peripheral nervous system. In *Notch*⁻ embryos all the ventral ectoderm makes neuroblasts while the dorsal ectoderm is little affected. In the developing imaginal disc, *Notch*⁻ clones behave in a comparable way: wherever there are usually closely packed bristles (like the anterior part of the thorax) **all** the *Notch*⁻ cells develop abnormally. In weak alleles, such as *Notch*ᵗˢ, extra bristles do form but in *Notch*⁻ the bristles do not differentiate at all and the cells make extra bristle precursors in abnormal places; these fail to mature properly. It seems likely that in *Notch*⁻, as in *shaggy*⁻, lateral inhibition fails, but that, in addition, normal *Notch* product is also needed for the later development of the bristles themselves. In the imaginal discs *Notch*⁻ clones are completely autonomous, meaning that in the area of the notum where overdispersed bristles are normally found **all** the *Notch*⁻ cells fail to form epidermal cells; surrounding *Notch*⁺ cells are not affected and no *Notch*⁻ cells are rescued by their neighbours. Embryos that are mosaic for *Notch*⁺ and *Notch*⁻ cells have been made and, again, the phenotype of *Notch*⁻ is cell autonomous. Thus in the ventral ectoderm, where neuroblasts are selected from amongst the cells, **all** *Notch*⁻ cells form neuroblasts, while in the dorsal ectoderm, where neuroblasts do not form, *Notch*⁻ cells form normal epidermis.

Cells defective in the *Delta* gene also fail in lateral inhibition and form masses of adventitious bristles in areas of the fly that normally bear bristles. However, the *Delta*⁻ cells at the extreme edges of the clones are **rescued** by the wildtype cells nearby and are able to make normal epidermis. As with the other two mutations, *shaggy*⁻ and *Notch*⁻, *Delta*⁻ clones are normal in areas where there are no bristles in the wildtype fly.

These results are in accord with the two-step model of neurogenesis: remember, the first step is the allocation of proneural clusters by the local expression in groups of cells of *scute* and other related genes. *Notch*, *Delta* and *shaggy* have no role here. The second step is lateral inhibition in which one cell of the proneural cluster is selected and here *Notch*, *Delta* and *shaggy* are required. It would follow from this hypothesis that lack of the *Notch* gene could cause no additional mischief if the *scute* functions were also missing. And this is so; if both *scute*$^+$ and *Notch*$^+$ are removed together the cells form a bristle-free, but otherwise normal epidermis.

And yet, on its own, *Notch*$^-$ transforms all epidermal cells into defective bristles; this also makes sense under the two-step model as Figure 7.5 explains. We compare wildtype and *Notch*$^-$; the model could apply both to bristles in the thorax and to neuroblasts in the embryo. The circles represent cells, some of which are dividing, the red ones expressing a *scute* RNA. When a cell becomes a bristle or a neuroblast it is shown in black. The figure suggests how, in the anterior notum where there are dense bristles, the mutant effects of *Notch*$^-$ might so deplete the reservoir of epidermal cells that gradually all the cells become incorporated into proneural clusters and all therefore develop into defunct bristles. By contrast, in the scutellum, there are only a few landmark bristles; the cells in the proneural clusters of these are damaged in *Notch*$^-$ and *Delta*$^-$ mutations, but the other cells are unaffected and divide and develop normally.

These and other experiments suggest that the three genes *shaggy*, *Notch* and *Delta* do not determine the pattern of *scute* expression and their mutations do not alter it. So, the overall distribution of bristly regions depends in part on the pattern of expression of the AS-C, while the selection of individual bristle cells depends on the three genes, and others. It is not easy to know how many cells are in a proneural cluster at the relevant time, but since the number of landmark bristles produced by mutations in *Delta* (to replace one wildtype bristle) is not more than seven, it is possible that the cluster of cells consists of a small group of cells — small enough so all its members could be in direct contact with each other. However, direct counts of the number of cells expressing *scute* RNA in a single proneural cluster in the imaginal disc gives a count of 20–30 cells — which is more than there are likely to be in any kind of direct contact.

How are the molecular components deployed? One smart approach uses genetic mosaics in which the cells of different genotype are differentially marked. The purpose is to study the selection of bristle cells from the proneural cluster and to see if the amount of *Notch* product influences the choice. If a bristle arises on the border between areas of cells of two genotypes it can be argued that it must have been selected from a proneural cluster of cells that was itself mosaic. In a control experiment there are clonal patches of cells that differ only in a marker mutation and in these, at the border of the clones, the bristles have an equal probability of originating from the clone or from the surround. In the experiment proper, one mosaic type is a mixture of *Notch*$^+$/*Notch*$^-$ (1 dose of *Notch*$^+$) and *Notch*$^+$/*Notch*$^+$ (2 doses) cells, another a mixture of *Notch*$^+$/*Notch*$^+$ (2 doses) and *Notch*$^+$/*Notch*$^+$/*Notch*$^+$ (3 doses). It is important to note cells of all these

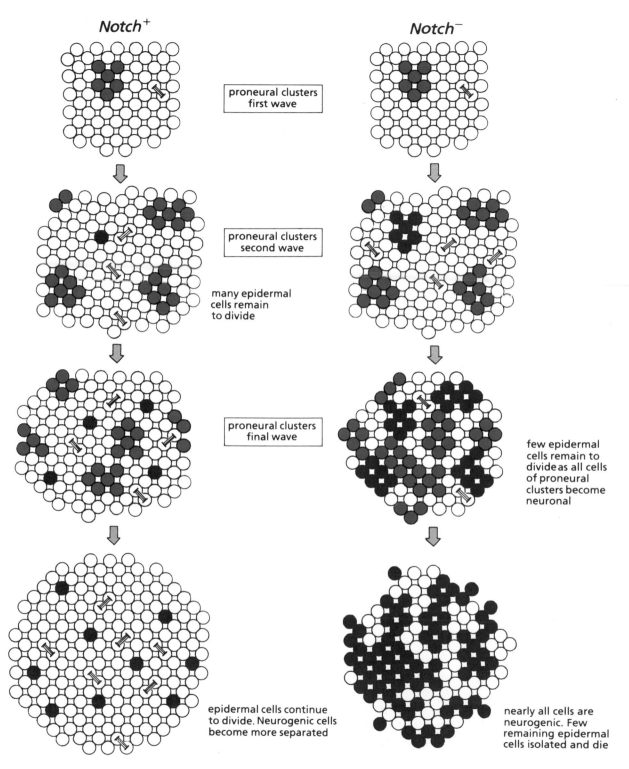

Notch⁺

Notch⁻

proneural clusters
first wave

proneural clusters
second wave

many epidermal
cells remain
to divide

proneural clusters
final wave

few epidermal
cells remain to
divide as all cells
of proneural
clusters become
neuronal

epidermal cells continue
to divide. Neurogenic cells
become more separated

nearly all cells are
neurogenic. Few
remaining epidermal
cells isolated and die

Figure 7.5 A model for neurogenesis in *Notch⁺* and *Notch⁻* epidermis.

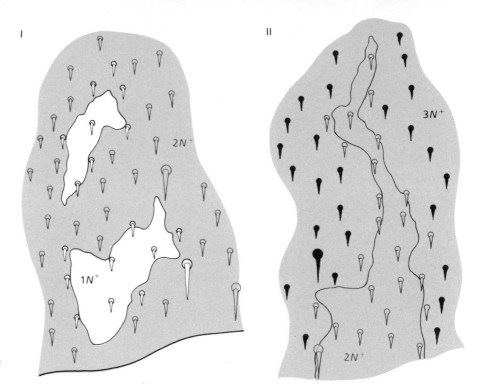

Figure 7.6 The effect of doses of *Notch* on bristle determination.

three genotypes are wildtype in the sense that they make virtually normal flies with completely normal bristle patterns on the thorax. Look at Figure 7.6: in I the small patches have 1 dose of *Notch*⁺ and make bristles (shown uncoloured). These patches form in a background of cells with 2 doses of *Notch*⁺, which are shown in red. All the bristles on the border of the patches, and some even a little outside, come from the cells with 1 dose of *Notch*⁺. In II the red cells and bristles carry 2 doses of *Notch*⁺ but now they are surrounded by cells with 3 doses (grey cuticle, black bristles). Now all the bristles at the border derive from the cells with 2 doses. The experiments are compelling; they show the behaviour of wildtype cells with 2 doses of *Notch*⁺ depends on the other cells in the cluster, proving that the fate of an individual cell is the result of a **comparison**, within the group of which it is a part. In other words the outcome is decided by competition between candidate cells, as suggested by the experiments described above (p. 165).

Lateral inhibition — the role of the *lin-12* and *Notch* genes

The *Notch* and *Delta* proteins show homology with the *lin-12* product, a gene in the nematode worm *Caenorhabditis elegans*. In the nematode, there are two neighbouring cells which always acquire two different fates. The two cells can be identified in the embryo and it is found that, in normal wildtype worms, it cannot be predicted which of the two cells will take up which fate; the allocation is stochastic. If either of the cells is killed, the other always develops one particular way and adopts what is therefore called the primary fate. In the absence of the *lin-12* product, both cells adopt the primary fate. There are dominant gain-of-function mutations of *lin-12* in which both cells adopt the secondary fate. Most tellingly, in genetic mosaics, the pair of cells may be of different genotype, one being *lin-12⁻* and the other *lin-12⁺*. In these mosaics the *lin-12⁻* cell **always** adopts the primary fate and the *lin-12⁺* cell always takes the secondary. The model for the role of *lin-12* relates to and refines other lateral inhibition models. It requires that the *lin-12* product is needed for reception of an inhibitory signal or is the receptor itself. Normally, the stochastic allocation of the two cells to primary and secondary fates will depend on some initial imbalance and one cell will begin to edge towards the primary fate. As it does so it produces more and more inhibitor which pushes the other cell more and more towards the secondary fate. The results from *lin-12⁺/lin-12⁻* mosaics demand this image of a gradual commitment: 'the bias in cell fate choice ... suggests that relative *lin-12* activity is assessed **prior** to cell fate decisions.' [11] Perhaps the production of inhibitor is linked, is consequent on, the absence of activation of the *lin-12* receptor. Thus, if a cell receives no inhibitor it will produce more inhibition and as a cell is inhibited it will produce less. The outcome will be the production of two disparate cell states, one completely inhibited and the other maximally inhibiting. Formally, the model is similar to that proposed for the blue-green alga (p. 161), but it places more emphasis on the receptor of the inhibitory signal, suggesting that it is a causal factor in the switch determining the cell state. The model predicts that if there were initially more *lin-12* receptor in one cell than in another, then the cell with least would pick up the least inhibitor and would edge towards the primary fate — it would then make more inhibitor, giving it a competitive advantage that would be decisive. This is exactly what was found in the mosaic experiment with the *Drosophila Notch* gene, a gene related to *lin-12*.

Notch, *Delta* and *lin-12* all encode transmembrane proteins with extracellular domains containing cysteine-rich repeats similar to those found in epidermal growth factor, a secreted protein, and low density lipoprotein (LDL) receptor which is an integral membrane protein. It is possible that both *Notch* and *lin-12* gene products are themselves receptors or involved in reception of inhibitory signals, and this is in accord with the cell autonomy shown by mutant cells in genetic mosaics. The non-autonomy of *Delta*, that is the rescue of *Delta⁻* cells at the periphery of a *Delta⁻* clone by the *Delta⁺* cells near to them, suggests that *Delta* protein may be part of the inhibitory signal.

The small number of cells in the proneural clusters and in the nematode cell systems suggests that direct contact may be needed for lateral inhibition. This is consistent with observations on *Notch* and *Delta* which hint they might be involved in cell contacts: if tissue culture cells express both *Notch* and *Delta* proteins, the two related molecules associate and copatch on the cell surface. Moreover, if *Notch*-expressing and separate *Delta*-expressing cells are mixed the two types of cells form coaggregates, although pure *Notch*-expressing cells will not aggregate. Possibly the *Notch* and *Delta* proteins combine to form a structural bond between cells, a bond that is necessary for, or involved in, an interchange of signals.

We have spent some pages on this topic, because, although the findings are not yet as clear as they might be, it is the elemental example of pattern formation — the spatially ordered generation of a new cell type.

What kind of bristle or neuroblast?

There are many kinds of sensory bristle; large, small, chemosensory or mechanosensory, and they are all placed appropriately. Most are clones, with each bristle descending from a single precursor cell, but some are not — for example the Keilin's organ, which straddles the parasegment border in the embryo and contains anterior as well as posterior cells, probably derives from two or more precursor cells. It is likely that several factors determine which type of sensillum will form. One is the timing of the selection process. In *Drosophila*, bristles that form early tend to be large landmark bristles; the smaller overdispersed bristles fitting in between them later on, so developmental context is important. Of primary importance will be which transcription factor genes are deployed — during neurogenesis a bewildering variety of transcription factors such as *ftz*, *eve* and *Krüppel* come on again in subsets of the neuroblasts and help determine cell type.

In bristle formation all cells of a proneural cluster have the potential to form the same type of sensillum; this is shown in *shaggy*⁻ clones when all the cells of a proneural cluster form bristles and they are all of the same structure and type as the bristle they have replaced. They all elicit the same behavioural response when stimulated, presumably because their ingrowing axons are similar and seek out the same sets of interneurones in the central nervous system.

There are genes that are required for the proper specification of sensilla-type; an example is the *cut* gene. There are two main sorts of sensilla in the peripheral nervous system of the embryo and larva. There are few of them and they are landmark organs — being invariant in position. External sensilla have various types of cuticular elements connected to a sensory dendrite. The others are called chordotonal organs; they are internal and span between two regions of the cuticle to monitor stretching. Both types of sensilla are made of four cells

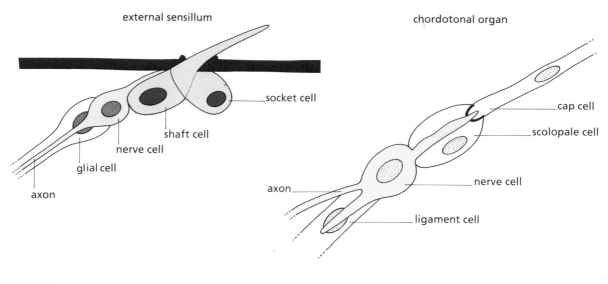

	external sensillum	chordotonal organ
cut⁺	external sensillum	chordotonal organ
cut⁻	transformed towards chordotonal organ	unaffected
heat shock cut	unaffected	transformed towards external sensillum

Figure 7.7 The effect of *cut⁻* and the ubiquitous expression of *cut* protein on two classes of organs in the peripheral nervous system of larvae.

(Figure 7.7). The *cut* protein is expressed in different amounts in the external sensilla nuclei (pink and red in Figure 7.7) but is absent from the chordotonal organs. In *cut⁻* embryos the external sensilla are transformed towards chordotonal organs which are themselves unaffected. When the *cut* gene is placed under heat shock control and universally expressed, the external sensilla are unaffected while the chordotonal organs are transformed towards external sensilla. These observations suggest that the *cut* product, a homeodomain protein, determines the identity of cells — thus, on a small scale, it shares some properties with selector genes such as the *Ubx* gene.

The lineage of bristle cells and neuroblasts

Once the bristle cell or neuroblast has been selected it undergoes a series of stereotyped cell divisions. Consider the bristle cell first. In *Drosophila* the bristle cell divisions have not been completely described, but they are better known in other flies and in other types of insects.

In the development of *Oncopeltus* bristles, the bristle mother cell also separates from the epidermal cells by rounding up and dividing in an unusual plane (epidermal cells always divide in the plane of the cuticle, these cells divide slightly obliquely). The daughter cell nearest the cuticle divides again to give two cells, one forms the socket and the other the shaft of the bristle. The daughter cell furthest from the cuticle also divides, in a plane perpendicular to that of the cuticle, and forms a nerve cell and a sheath cell; this division is unusual, it is asymmetric, generating a larger and a smaller daughter.

Some bristles have several nerve cells and these are generated by extra divisions of the nerve precursors; some bristles are uninnervated, in these the divisions are gone through but the presumptive nerve and sheath cells die. More than 50 years ago Stossberg observed the special divisions of the scale mother cells in a moth and made delicate drawings of them. The first division of an erstwhile epidermal cell is vertical and asymmetric, giving an internal smaller daughter that degenerates. The outer cell divides again slightly obliquely to give an external cell that makes the socket, and an internal cell that makes the scale (Figure 7.8). The **impression** gained from these stereotyped patterns of divisions in different systems is of a mechanism intimately linked to the divisions themselves — some kind of programme that is read out through the divisions and which does not depend on interaction between the cells that are generated.

This type of 'cell lineage' model has always proved very attractive to the scientist; it is rigid, digital, it works like a computer — unlike most of animal development in which it is hard to detect the hidden order in what appears to be chaos, but cannot be. One class of theories

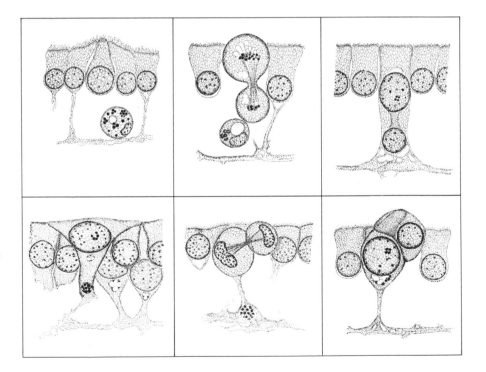

Figure 7.8 Stossberg's original drawings of scale development in a moth, *Ephestia*. The drawings show the two divisions which generate the two cells that make the scale and its socket. One daughter cell of the first division degenerates.

involves the unequal segregation at mitosis of chromosomal or at least nuclear material — so that the nascent nuclei would be different from the beginning. Another class of theory emphasises the cytoplasm — perhaps the two cells have different cytoplasmic components, again due to unequal segregation during cytoplasmic cleavage. Many experiments in embryology argue in general against the former view, for example the capacity of differentiated nuclei from the gut of frog tadpoles to go through development again, if they are transplanted into oocytes, emphasises that nuclei do not change much, or irrevocably, as development proceeds. There is one particularly relevant example from insects; when embryonic neuroblasts divide they generate two very different daughters. The smaller cell becomes a ganglion mother cell and divides once more to produce two neurones, while the larger cell resembles its mother and again divides asymmetrically to produce a larger cell like itself and a ganglion mother cell. The larger cell buds off a chain of smaller ganglion cells in reiterated divisions. If these cells from a grasshopper are cultured *in vitro* they continue to produce ganglion cells, and this process carries on unaffected if, at anaphase of mitosis, the mitotic spindle is rotated, swapping the nuclear material between the two nascent cells. The differences between the two cells in how they behave and in their developmental potential are not therefore in the nuclear material but, presumably, in the cytoplasm.

There is one organism that develops almost entirely like some giant bristle, and that is the nematode *C. elegans*. There is a complex stereotyped lineage tree that generates the 959 cells of the main body, all the way from the zygote. This too is attractive, apparently as an example of a different mode of development that is more tractable. But, as we have seen, it is increasingly becoming clear that even this regimentation of lineage is the ultimate outcome of interactions between cells and other rather 'messy' mechanisms that typify most developmental processes. In some innervated moth scales, if the nerve cells are killed soon after they are formed, the scale and socket cells do not develop properly. In the nematode, the elimination of cells is sometimes followed by change of fate in the remainder; both these examples speaking for cell interaction as a cause of the cells becoming different in the first place. As in most areas of developmental biology, the truth will no doubt include elements of both conflicting theories; although there must be many cell interactions, it may well be that the polarity of single cells is made use of to divide the cytoplasm up unequally, and that the resultant size difference can be instrumental in directing the two daughter cells down divergent paths without much or any 'conversation' between them.

To make progress on this topic in *Drosophila*, we need a thorough genetic and molecular analysis of the process of bristle development along the lines being pursued in the eye (Chapter 8) and this has only just begun. There are a number of mutations that interfere with the allocation of cells within the lineage tree of small sensory organs such as bristles. *Delta⁻* mutations (in mitotic recombination clones) cause misallocation of bristle cells, so in those clones there are frequently double-socketed or double-shafted bristles. Likewise, if the *Notch* gene product is removed or reduced during the time when the bristle mother cell is dividing (this is done by giving a heat pulse to pupae carrying a *Notch^{ts}* allele) the allocation of the daughter cells is affected and only neurones are produced. Given the current model of the *Notch* protein as being involved in reception of an intercellular signal (p. 170), this observation suggests that intercellular signals act in the allocation of the daughter cells of the bristle mother cell.

As in other systems, transcription factors might be expected to fix cell identity after the initial cell allocation. Indeed, a gene called *numb*, whose sequence and the nuclear localisation of the protein suggest a transcription factor, is required for cell allocation in the larval peripheral nervous system. In the absence of the *numb* gene, there is a switch of cell identity in two of the four cells of the sensillum, the neurone and sheath cell being transformed into additional socket and shaft cells.

There are two further points about bristle patterns that are educational: first, in the leg of *Drosophila*, the bristles are each associated with a little bract that is proximal to the socket and points distally (see Figure 6.1, p. 137). Analysis of clones marked with such a mutation as *yellow⁻* shows that the bract is not a descendant of the bristle precursor cell and therefore must be the result of recruitment of a normal epidermal cell by the bristle — intimations of the development of pattern in the eye (Chapter 8).

The second point concerns cell movement in the formation of bristle patterns. In *Oncopeltus*, the three cells of the hair primordium move about and line up like policemen after they have formed, their orientation predicting that of the bristle outgrowth and illustrating their sensitivity to the segmental gradient (p. 140). In moths, the scale cells, originally relatively scattered, line up in serried ranks. In the *Drosophila* wing margin there are three rows of tightly packed and precisely arranged bristles; it is probable that they shuffle into position after the cells have been allocated. The evidence for this is that small clones of marked cells become fragmented at the wing margin and often mark bristles that are not contiguous. It has even been suggested that as the bristle cells shuffle, they might mix up the descendants of single mother cells — so that the socket cell arising from one mother cell might partner the shaft cell coming from another, and thus single bristle organs would not always be clones. If this were correct — and whether it is is unclear — the triple row on the wing margin would form by a kind of self assembly and would share some features with the eye (Chapter 8). More evidence for relative movement between bristles and epidermal cells is found in the *Drosophila* abdomen; clones usually contain both epidermal cells and bristles and these quite often become separated, the bristle belonging to the clone becoming surrounded by other cells or, conversely, the clone may surround bristles that originated outside it (see Figure 5.3, p. 112). The explanation might be that, during metamorphosis, as the adult epidermal cells migrate amongst the moribund larval cells, they could leave the newly formed bristle cells behind.

Further reading

REVIEWS

Ghysen, A. and Dambly-Chaudiere, C. (1989) Genesis of the *Drosophila* peripheral nervous system. *Trends Genet.* **5**: 251–255.

Greenspan, R.J. (1990) The *Notch* gene, adhesion, and developmental fate in the *Drosophila* embryo. *New Biologist* **2**: 595–600.

Greenwald, I.S. (1989) Cell–cell interactions that specify certain cell fates in *C. elegans* development. *Trends Genet.* **8**: 237–241.

Horvitz, R.H. and Sternberg, P.W. (1991) Multiple intercellular signalling systems control the development of the *Caenorhabditis elegans* vulva. *Nature* **351**: 535–541.

Simpson, P. (1990) Lateral inhibition and the development of the sensory bristles of the adult peripheral nervous system of *Drosophila*. *Development* **109**: 509–519.

Stern, C. (1968) *Genetic Mosaics and Other Essays*. Harvard University Press, Boston.

Wigglesworth, V.B. (1959) *The Control of Growth and Form: A Study of the Epidermal Cell in an Insect*. Cornell University Press.

SELECTED PAPERS

Achaete-scute

Cubas, P., de Celis, J.-F., Campuzano, S. and Modolell, J. (1991) Proneural clusters of *achaete/scute* expression and the generation of sensory organs in the *Drosophila* imaginal wing disc. *Genes Dev.* **5**: 996–1008.

Rodríguez, I., Hernández, R., Modolell, J. and Ruiz-Gómez, M. (1990) Competence to develop sensory organs is temporally and spatially regulated in *Drosophila* epidermal primordia. *EMBO J.* **9**: 3583–3592 (1990).

Stern, C. (1954) Two or three bristles. *Am. Sci.* **42**: 213–247.

Villares, R. and Cabrera, C.V. (1987) The *achaete-scute* gene complex of *D. melanogaster*: conserved domains in a subset of genes required for neurogenesis and their homology to *myc*. *Cell* **50**: 415–424.

Anabaena

Wilcox, M., Mitchison, G.J. and Smith, R.J. (1973) Pattern formation in the blue-green algae *Anabaena*: I Basic mechanisms. *J. Cell Sci.* **12**: 707–723.

Bristle spacing in other insects

Lawrence, P.A. and Hayward, P. (1971) The development of a simple pattern: spaced hairs on *Oncopeltus fasciatus*. *J. Cell Sci.* **8**: 513–524.

Wigglesworth, V.B. (1940) Local and general factors in the development of 'pattern' in *Rhodnius prolixus* Hemiptera. *J. Exp. Biol.* **17**: 180–200.

cut

Blochlinger, K., Jan, L.Y. and Jan, Y.N. (1991) Transformation of sensory organ identity by ectopic expression of Cut in *Drosophila. Genes Dev.* **5**: 1124–1135.

daughterless

Caudy, M., Vässin, H., Brand, M., Tuma, R., Jan, L.Y. and Jan, Y.N. (1988) *Daughterless,* a *Drosophila* gene essential for both neurogenesis and sex determination, has sequence similarities to *myc* and the *achaete-scute* complex. *Cell* **55**: 1061–1067.

lin-12

Seydoux, G. and Greenwald, I. (1989) Cell autonomy of *lin-12* function in a cell fate decision in *C. elegans. Cell* **57**: 1237–1245.

Neuroblast development in the grasshopper

Carlson, J.G. (1952) Microdissection studies of the dividing neuroblast of the grasshopper, *Chortophaga viridifasciata* (de Geer). *Chromosoma* **5**: 199–220.

Notch

Fehon, R.G., Kooh, P.J., Rebay, I., Regan, C.L., Xu, T., Muskavitch, M.A.T. and Artavanis-Tsakonas, S. (1990) Molecular interactions between the protein products of the neurogenic loci *Notch* and *Delta*, two EGF-homologous genes in Drosophila. *Cell* **61**: 523–534.

Hartenstein, V. and Posakony, J.W. (1990) A dual function of the *Notch* gene in *Drosophila* sensillum development. *Dev. Biol.* **142**: 13–30.

Heitzler, P. and Simpson, P. (1991) The choice of cell fate in the epidermis of Drosophila. *Cell* **64**: 1083–1092.

Hoppe, P.E. and Greenspan, R.J. (1990) The *Notch* locus of *Drosophila* is required in epidermal cells for epidermal development. *Development* **109**: 875–885.

Simpson, P. (1990) *Notch* and the choice of cell fate in *Drosophila* neuroepithelium. *Trends Genet.* **6**: 343–345.

numb

Uemura, T., Shepherd, S., Ackerman, L., Jan, L.Y. and Jan, Y.N. (1989) *numb*, a gene required in determination of cell fate during sensory organ formation in *Drosophila* embryos. *Cell* **58**: 349–360.

Scale development

Stossberg, M. (1938) Die Zellvorgänge bei der Entwicklung der Flügelschuppen von *Ephestia kühniella* Z. *Z. Morph. Ökol. Tiere.* **34**: 173–206.

shaggy

Bourouis, M., Heitzler, P., El Messal, M. and Simpson, P. (1989) Mutant *Drosophila* embryos in which all cells adopt a neural fate. *Nature* **341**: 442–444.

For details, see above.

Figure 7.1 After Wigglesworth (1940).

Figure 7.2 See Lawrence and Hayward (1971).

Figure 7.3 See Wilcox *et al.* (1973).

Figure 7.4 After Stern (1954).

Figure 7.5 After Simpson (1990).

Figure 7.6 After Heitzler and Simpson (1991).

Figure 7.7 See Blochlinger *et al.* (1991).

Figure 7.8 From Stossberg (1938).

Plate 7.1 Photographs courtesy of R.I. Vane-Wright and B. D'Abrera.

Plate 7.2 See Cubas *et al.* (1991). Photographs courtesy of J. Modolell.

8 The eye

THE COMPOUND EYE consists of a crystal-like array of ordered sets of cells and is, therefore, an excellent model system to study small-scale patterns. Each ommatidium is built up progressively as a single unit, and, at each step, specific genes and interactions work to allocate cells from the epithelium. Eventually nearly all the cells fit into the lattice and the remainder die.

Throughout this book it has been clear that, for most of the processes studied genetically, only some of the genes have been identified. Attempting to study a process with only a subset of the genes involved carries a risk: like trying to understand how a car engine works from a few pieces — say, a piston, the main gasket and the bolts that attach it to the chassis. In the embryo, a systematic attempt has been made to find the genes involved in early patterning but, because of maternal rescue (Box 1.3, p. 8), this cannot be completely successful. In the development of the eye an attempt is being made to identify **all** the genes and this has the advantage that the eye is dispensable, meaning that completely eyeless (and *eyeless⁻*) flies are viable. It follows that mutations in genes whose **only** role is in the construction of the eye will be viable, too. There will be genes needed for the eye as well as for other parts and their role in the eye can be more difficult to assess as mutations in them will often be lethal. However, clones of cells that lack these essential genes can still be studied in the eye.

The eye has many advantages in addition to the genetic ones; it is above all a beautiful structure in which the full capacities of developmental processes are displayed. The eye is made of some 750 repeating units, each being a little eye or ommatidium. Each has a lens made by four cells that focuses the light down a channel and eight photoreceptor cells (R1−R8) arranged in a precise pattern within the channel. Figure 8.1 shows a section through the eye as seen down a light microscope. Many ommatidia can be seen; the equator is arrowed — it is a mirror plane of symmetry which can be picked out by comparing the arrangement of the photoreceptor cells R1−R7 that appear as a pattern of dots in each ommatidium. Surrounding the lenses are two primary pigment cells; secondary and tertiary pigment cells enclose and insulate the light channel of each ommatidium. At the base of the retina there is a network of pigmented subretinal cells. The eye is also equipped with

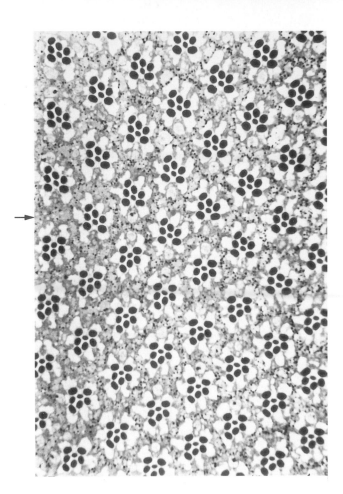

Figure 8.1 Section through the normal eye of *Drosophila* to show the arrangement of the ommatidia and the photoreceptor cells. Note the equator (arrow).

mechanosensory bristles, and these are placed between the ommatidia — each bristle is made by four cells. The structure of an ommatidium is summarised in Figure 8.2 which shows a longitudinal section and a series of cross-sections. Note that the R8 cell underlies the R7 cell, so it is only visible in deeper sections. All the ommatidia are packed together so precisely that the whole has been called 'a neurocrystalline lattice' [12]; generally there are no missing cells, nor are there extra cells that do not fit in.

This orderliness extends to the way the compound eye is wired, for, under the convex surface of the eye, there are two layers of the brain containing an array of neurones. The central photoreceptor cells, R7 and R8, send axons directly through the lamina to the underlying medulla, while the other photoreceptor cells connect to pairs of cells in the lamina with a precision reminiscent of a Swiss telephone exchange. The effect of the wiring is to ensure that the set of six photoreceptor cells, which come from six different adjacent ommatidia and which look at one point in space, are all connected to the same pair of laminal neurones. This unites an initially fragmented image. A careful study of

181 THE EYE

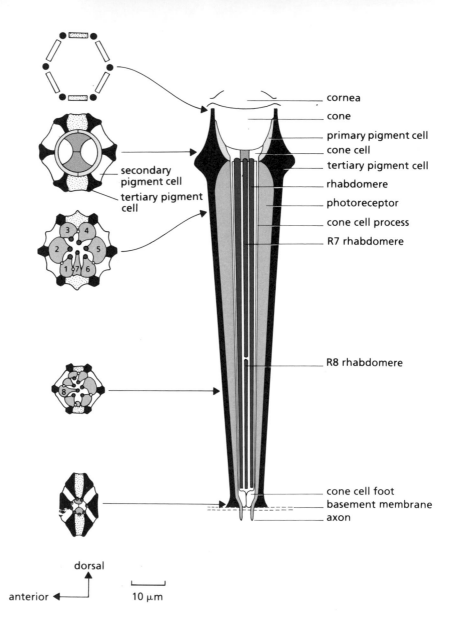

Figure 8.2 The structure of an ommatidium. One longitudinal section and five cross-sections are shown.

dorsal

anterior

10 μm

the accuracy of this process, which was done in the blowfly, *Calliphora*, showed that very few mistakes are made. Of 650 neurones traced none went to the wrong cells in the lamina. Many of these neurones came from the equator (Figure 8.1) where the situation is complicated by the mirror-image reversal; here as many as eight different photoreceptor cells, from eight different ommatidia, may look at one point in the visual field. One group is shown in red in Figure 8.3. Several such sets were traced and it was found that in every case the eight cells were connected together unerringly to a single cell in the lamina.

It is not known how the correct connections are achieved, only that the development of laminal order is dependent on the ommatidial array itself. If clones of mutant cells are made in the eye that disrupt

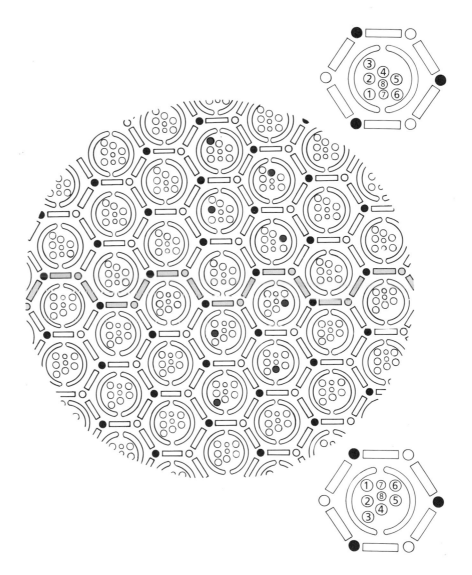

Figure 8.3 The projection from the ommatidial array. Near the equator (shaded) there are up to eight cells (one set is shown in red) which see one point in space, and send axons to the same cell in the underlying lamina of the brain.

that array, then the normal wildtype lamina underneath develops with a disrupted pattern. The disrupted region in the lamina correlates in area and position with the overlying clone. Even more persuasively, the experiment can be done in reverse: there is a dominant mutation, *Glued*, which makes a mess of both the eye and the lamina. The mutant ommatidia are higgledy piggledy, with variable numbers of photoreceptor cells, while the lamina is packed with disorganised nerve fibres instead of the normal array of special cells that receive input from the ommatidia. By mitotic recombination, the deleterious *Glued* allele can be removed from clones in the eye to give a patch of wildtype ommatidia that are arranged in a normal pattern. These normal ommatidia send their axons into the lamina and, there, the mutant *Glued* cells receive them. Surprisingly, the lamina becomes organised in that region and a normal array of cells is made. Some degree of order is then propagated even further into the medulla of the brain. The mechanism for these interactions is unknown: one would guess that the sequence of axon ingrowth is important — the R8 axon grows in first, and the others follow in a spatial and temporal order. The 750 reiterated units (approximately 16 000 cells), the specific cell types and the precision make the eye a good system to analyse at the cellular and molecular levels and with the genetic approach.

Because it is so accurately made, it used to be blithely assumed that the mechanism for generation of the cells must be a lineage tree — each ommatidium being a clone descending from a single ommatidial mother cell. This idea was accepted for decades, but not tested; when it was tested it was found to be wrong. Two methods of test have been used; one is to look along the border of a gynandromorph (p. 11) in which the female cells are wildtype with red pigment while male cells are *white⁻* and colourless. What is found is that the male and female cells are mingled along the border and individual ommatidia are of mixed genotype — therefore the ommatidia cannot be clones. However there is still the possibility that the different cell types might be

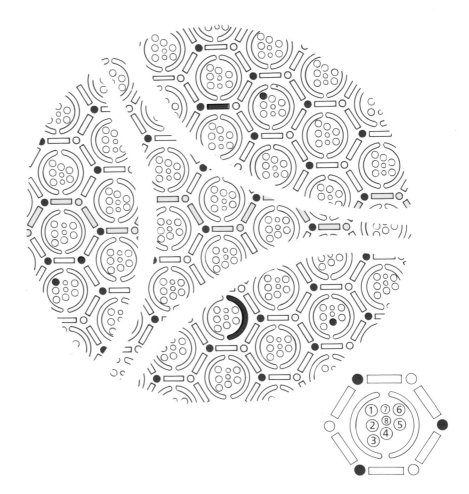

Figure 8.4 Three separate *white⁺* clones, in three different *white⁻* eyes. Each clone consists of two cells; they illustrate that different kinds of cells can be sisters.

generated by cell lineage, for example there might be a class of stem cells that give rise to only secondary pigment cells, and then these related cells would migrate between different ommatidia to take up their positions, and this could produce ommatidia of jumbled genotypes. This possibility was eliminated by making very small marked clones at the end of eye development. The most useful clones consist of only two sister cells, because if there are any lineage restrictions these clones should reveal them. For example, if the photoreceptor cells diverge in lineage from pigment cells, then it would not be possible for a pigment cell to have a photoreceptor cell as a sister. The results show that any pair of cell types can be sister cells. Look at the illustrations in Figure 8.4 and Plate 8.1 which show small *white*⁺ clones. Each consists of two cells; probably the labelled cells are sister cells; 'probably' because there is always the chance that other cells belonging to the clone could have died or formed another cell, such as a cone cell, that is not scored. Plate 8.1 shows a clone that labels only one pigment cell and one R1 cell.

Having eliminated the role of cell lineage, we must embrace the truth that other methods can produce a 'neurocrystal' and to ask what they are. There is evidence that each ommatidium is made as a unit, because even a single one can form in isolation; in *Ellipse* eyes (Figure 8.5) there are fields of improperly differentiated cells interspersed with an occasional complete ommatidium. This is a very helpful observation because it directs the analysis away from the idea that the eye could grow just like a crystal with 'cosy corners', niches into which undetermined cells would fit. In that model, all cells would be equivalent in the sense that they would all become lodged into the growing face of the crystal and none would lead the process. Instead we have to think of ommatidia starting life as separate units and fitting together later.

The eye does not develop all at once; the posterior ommatidia are formed first and the anterior last: as they mature in succession a

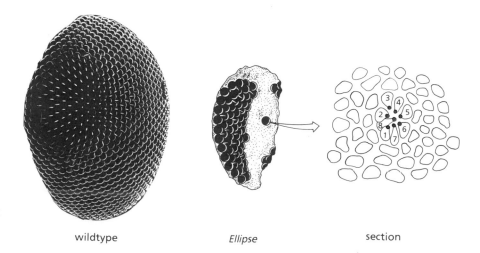

Figure 8.5 In eyes mutant for *Ellipse*, single isolated ommatidia can be found.

wildtype   *Ellipse*   section

furrow, like the front of a wave, sweeps slowly across the eye imaginal disc. This helps analysis because one eye disc will contain maturing ommatidia in all states, earliest stages in the furrow and later ones behind it. Figure 8.6 shows an eye-antennal disc with the developing ommatidia shown up by an antibody. The furrow is moving from right to left, that is from posterior to anterior. The process of ommatidial development can be broken down into a succession of steps.

First, groups of cells appear, separated from one another as in a spacing pattern (Chapter 7). Initially each is a small group of five or six cells that partially encloses one or two cells. These central cells have no known fate in the mature ommatidium and are therefore called mystery cells. Within the group the first to differentiate is the cell that will become R8 (red). The process of selection of these groups shares features with other spacing patterns, for example reduction in the

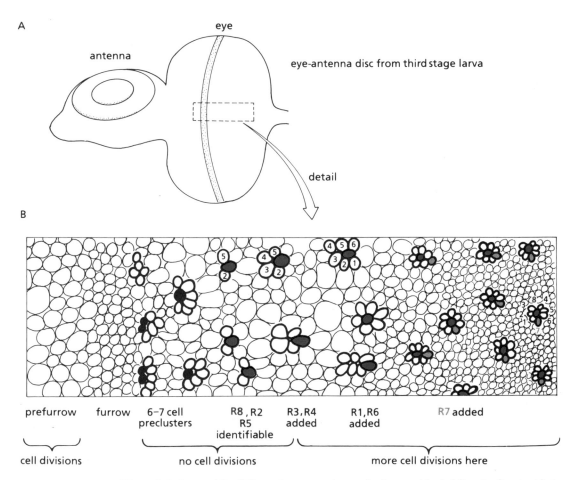

Figure 8.6 Ommatidia differentiate posterior to the furrow (shaded line in disc in A). In B, the appearance of the **apical** surfaces of the cells is indicated. The furrow moves steadily anteriorwards so that the most mature stages are at the posterior (on the right). Mystery cells in black, R7 in red.

amount of *Notch* protein causes the specification of many extra photo-receptor cells which develop at the expense of the pigment and cone cells (see p. 182). However, in clones of cells lacking the achaete-scute complex (p. 162) the ommatidia are normal, although the eye bristles do not form.

Second, two cells neighbouring the R8 cell begin to differentiate; these will become the R2 and the R5 cells and, outside these, the presumptive R3 and R4 cells can be detected. Soon this little group of cells rearranges itself into a circle, casting out the mystery cells as it does so. The result is a central R8 cell with R2, R3, R4 and R5 placed around; the apical contacts between these cells are relatively stereotyped so that the cells can be identified in electron microscope sections.

Third, two cells adjacent to the cluster, and sitting in precise positions relative to the determined cells, become allocated to R1 and R6 and, finally, a third joins them as R7. The four cone cells are also identifiable in the clusters; the pigment cells fit in and around them at a later stage. Initially the clusters all have the same orientation, pointing posteriorly (just like the polarised group of cells that form *Oncopeltus* bristles (p. 176)) but later they rotate in one direction in the dorsal half of the eye and the other in the ventral half, generating the two forms of ommatidia which meet at a mirror plane at the equator of the eye (Figure 8.1). These stages can all be seen in a single disc; a drawing of the process is given in Figure 8.6B.

The little bristles form quite late and do not follow the same anteroposterior sequence as the rest of the eye; instead the process starts at the centre and spreads outwards. They fit into the lattice between the ommatidia and consist of four cells that are presumably the descendants of a single bristle mother cell (as in other bristles, p. 172). As these processes are occurring, the undetermined cells continue to divide, and occasionally, mainly at the end of the process, excess cells die. The total proportion of cells dying has been estimated at a little over 10%.

The genetic approach to this problem has been effective and is still accelerating. The strategy is to identify genes required for ommatidial formation from mutations, to use clones of cells that lack the gene to find out where the gene is required, and then to clone the gene, deduce the nature of its product and determine when and in which cells it is expressed. Cloning the gene can be followed by expressing the gene in inappropriate cells or at unusual times and studying effects on the pattern. I will take the *sevenless* gene as an example.

The *sevenless* gene was found in a systemic screen for mutations with defective vision; the ommatidia lack R7 and the cell which would have made R7 instead makes a cone cell. The ommatidia develop with the usual four cone cells and it is not known what happens to the cell which, in the wildtype, would have made that cone cell. Perhaps it just

Figure 8.7 A *sevenless⁻ white⁻* clone in a *white⁺* (red) background. Ommatidia lacking the R7 cell have a different arrangement of cells R1−R6 (bottom, right, and in clone).

sevenless⁺

sevenless⁻

remains in the pool of unallocated cells. The requirement for the *sevenless* gene is autonomous: this is shown by making clones of cells that are *white⁻ sevenless⁻* in a wildtype background and analysing ommatidia that are mixtures of mutant and wildtype cells. It is found that when an ommatidium contains an R7 cell, that cell is **always** *white⁺ sevenless⁺* and this is true even when all the other cells in the ommatidium are mutant. No *sevenless⁻* R7 cell is ever formed, even if all the other cells in the ommatidium are wildtype. An example of a mosaic eye is shown in Figure 8.7. An arrow indicates an ommatidium that shows the normal wildtype pattern of R cells — even though, apart from the *sevenless⁺* R7 cell, it is entirely of mutant *sevenless⁻* genotype. Clearly the *sevenless* gene is required in the R7 cell and in the R7 cell alone.

The *sevenless* gene has been cloned and been found to encode a transmembrane receptor-like protein with an intracellular tyrosine kinase. As would be expected, the gene is expressed in the R7 cell; it is also found in other cells nearby but not in R8, R2 and R5. This is seen in the electron microscope if the *sevenless* antigen is stained (Figure 8.8). The picture shows the apices of the cells and it is clear that the protein appears in those cell membrane regions of R7, R3 and R4 exactly where they contact R8. This suggests they are forming patches as and because they bind to some ligand on the R8 cell. Inside the three cells that express the *sevenless* antigen, multivesicular bodies can be seen that are densely stained (arrow). This expression of *sevenless* in the R3 and R4 cells might suggest some function in those cells, but as far as is known, the R3 and R4 cells are normal in *sevenless⁻* eyes. Ubiquitous expression of *sevenless* under heat shock control has no effect (except to rescue *sevenless⁻* eyes) suggesting that it is the location of the ligand that is important in the wildtype, and that extra receptor in unusual places has no consequence, presumably because it is not activated. If very large amounts of *sevenless* protein are produced in the eye (and this can be done by having as many as eight copies of the wildtype gene, or by driving gene expression with an especially active promoter) then the level of tyrosine kinase function can reach a point where it triggers downstream effects in the cells. In this artificial situation, some of the photoreceptor cells and the cone cells are transformed and now make extra photoreceptor cells of the R7 type.

One might therefore expect another 'sevenless' gene whose role is in the R8 cell to produce the ligand for the *sevenless* product. One candidate gene has been found and has been called *bride-of-sevenless* (*boss*). *boss⁻* eyes are normal except that they lack the R7 cell. Genetic mosaics show that, in order to obtain a normal R7 cell, *boss⁺* is

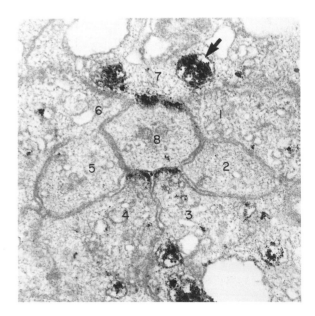

Figure 8.8 The localisation of *sevenless* antigen in the cell apices of a developing ommatidium. Numbers show different R cells. Arrow indicates a multivesicular body.

required in only the R8 cell of that ommatidium. The requirement for *boss*⁺ is therefore non-autonomous (because *boss* affects the R7 cell but is not needed in the R7 cell itself). Look at Figure 8.9 which shows a section through part of a pigmented *white*⁺ eye and includes part of a clone of *white*⁻ *boss*⁻ cells. Ommatidia show the sevenless-like phenotype in the white territory and the normal phenotype in the *white*⁺ region. Near the border, mosaic ommatidia are found. One ommatidium illustrates the non-autonomy of the *boss* function; not only is the pattern of the R1−R7 cells normal, but also the R7 cell itself is genetically *white*⁻ and therefore must be *boss*⁻. In this ommatidium, the R8 cell is *boss*⁺. The *boss* gene encodes a receptor-like membrane protein with a large extracellular domain and such a protein could, in principle, interact with the *sevenless* receptor, and there is good evidence that it does: *boss* protein is expressed in the R8 cell at the right time; also *boss* and *sevenless* proteins will cause cell aggregation when expressed in tissue culture cells (p. 172).

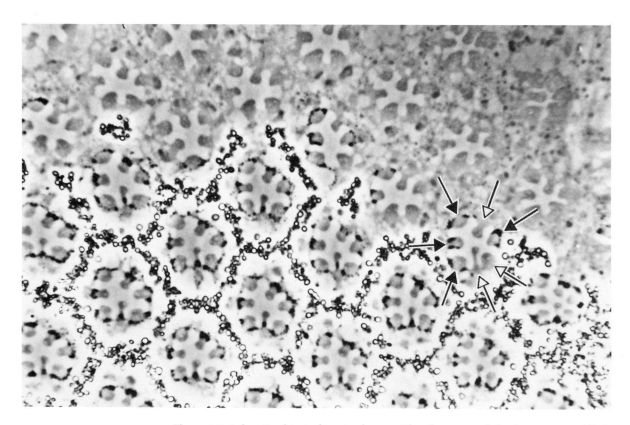

Figure 8.9 A *boss*⁻ *white*⁻ clone in the eye. The phenotype of the *boss*⁻ ommatidia is like that of *sevenless*⁻, but the *boss* gene is required in R8 and not in R7. In one mosaic ommatidium all the cells are present, even though the R4, R6 and R7 cells are *white*⁻ *boss*⁻ (open arrowheads) while the R1, R2, R3 and R5 are *white*⁺ and *boss*⁺ (closed arrowheads).

By extrapolation from the embryo, one would also expect selector genes to be active in the eye, not only those that determined the development of the whole compartment in the embryo (such as the *Deformed* gene which is expressed in the eye imaginal disc) but also other genes encoding transcription factors. Again, there is one good and instructive example and that is the *rough* gene. *rough⁻* flies are viable and their eyes are a disorganised mixture of cells: genetic mosaics show that the *rough* gene is required only in the R2 and R5 cells and is needed for the proper development of R3 and R4; it seems likely that the other effects on developing ommatidia stem indirectly from this, for the chain of normal events will be broken if R3 and R4 fail to differentiate. The *rough* gene encodes a transcription factor with a homeodomain and therefore it is no surprise that it is found in the nucleus. Any role of *rough* in cell interactions between R2/R5 and R3/R4 can only be indirect, since these interactions would presumably be mediated by secreted factors or membrane bound ligands; the genes involved in these steps are not yet identified. The most satisfying interpretation of the *rough⁻* phenotype is that, in the absence of the *rough* product, the R2/R5 cells are no longer quite themselves and although they look normal they do not do all the things they should, such as influence the presumptive R3 and R4 cells, perhaps by producing a special ligand. Like many genes we have come across, the *rough* product is expressed in a complex and changing pattern that is not easy to understand. Indeed it may never be necessary to understand it; what is important is that *rough* has a role in R2 and R5 and, apparently, nowhere else — the transcription that is seen elsewhere may be inconsequential. Nevertheless, if the *rough* gene is expressed strongly in extra cells by putting it under heat shock control it has severe effects on the eye. If *rough* is put under the direction of the promoter region of the *sevenless* gene (so that it will be expressed in the R7 cell, where it is normally absent) the R7 cell is changed and now resembles a cell in the R1−R6 class.

As developing R2 and R5 cells are distinguished by the expression of the *rough* gene, so are R3, R4, R1 and R6 characterised by expression of, and requirement for, the *seven-up* gene. Clones of *seven-up⁻* cells in the eye are normal in R8, R2, R5 and R7 but have small rhabdomeres in R3, R4, R1 and R6 cells. The *seven-up* gene encodes a zinc finger gene of the steroid receptor type and presumably acts as a transcription factor required for the proper differentiation of these four R cells. As would be expected from the order of cell allocation in the ommatidium, *rough* acts upstream of *seven-up*; in *rough⁻* eyes the *seven-up* gene becomes expressed in R2 and R5, causing them to be transformed to, or towards, the R3/R4/R1/R6 type.

From all this work a picture of eye formation is beginning to emerge and it presents most of the problems facing the developmental biologist in microcosm; it will take much more to make that picture complete but there are already valuable lessons about building patterns. First, there are the large-scale effects which place and polarise the eye as a whole — these are the early acting genes such as those of the Antennapedia complex that specify development of the whole para-segment. There are the segmental gradients which presumably provide overall polarity to the responding cells, and these determine how the group of cells is oriented as well as in which part of the imaginal disc ommatidia will form. Then there are the spacing mechanisms that choose, from a sheet of more homogeneous cells, the clusters that will form the ommatidia. Genes used in other spacing patterns have a role here. Then there are small scale effects; the allocation of nearby cells to form the R2 and R5 and the chain of events leading to allocation of the remaining cell types including R7. The *sevenless* gene is best understood; it encodes a receptor protein suggesting that a ligand from a neighbour cell, a short-range signal, coming from R8, is the trigger that leads that cell to differentiate into R7. One imagines a whole series of these neighbour interactions in which numbers of ligands and receptors (some unique to the eye and some not) are used. There are the other cells, the cone cells, the primary and secondary pigment cells that must fit in here somewhere — do they depend on cell-specific signals too? Then there is the little bristle, presumably a clone. Remember, too, the role of small-scale cell movements and changes of shape. The cells move about a little and pack together as intricately as in a Rubik cube as they construct the eye — there is a great deal of cell biology and mystery there.

Further reading

REVIEWS

Ready, D.F. (1989) A multifaceted approach to neural development. *Trends Neurosci.* **12**: 102–110.
Rubin, G.M. (1991) Signal transduction and the fate of the R7 photoreceptor in *Drosophila*. *Trends Genet.* In press.
Tomlinson, A. (1988) Cellular interactions in the developing *Drosophila* eye. *Development* **104**: 183–193.

SELECTED PAPERS

bride-of-sevenless

Krämer, H., Cagan, R.L. and Zipursky, S.L. (1991) Interaction of *bride of sevenless* membrane-bound ligand and the *sevenless* tyrosine-kinase receptor. *Nature* **352**: 207–212.

Cell lineage

Lawrence, P.A. and Green, S.M. (1979) Cell lineage in the developing retina of the *Drosophila*. *Dev. Biol.* **71**: 142–152.

Ready, D.F., Hanson, T.E. and Benzer, S. (1976) Development of the *Drosophila* retina, a neurocrystalline lattice. *Dev. Biol.* **53**: 217–240.

Ellipse

Baker, N.E., and Rubin, G.M. (1989) Effect on eye development of dominant mutations in *Drosophila* homologue of the EGF receptor. *Nature* **340**: 150–153.

Eye–brain connections

Horridge, G.A. and Meinertzhagen, I.A. (1970) The accuracy of the patterns of connexions of the first- and second-order neurons of the visual system of *Calliphora*. *Proc. R. Soc. Lond. B* **175**: 69–82.

Meyerowitz, E. and Kankel, D. (1978) A genetic analysis of visual system development in *Drosophila melanogaster*. *Dev. Biol.* **62**: 112–142.

rough

Tomlinson, A., Kimmel, B.E. and Rubin, G.M. (1988) *rough*, a *Drosophila* homeobox gene required in photoreceptors R2 and R5 for inductive interactions in the developing eye. *Cell* **55**: 771–784.

Sequence of cell allocation

Tomlinson, A. and Ready, D.F. (1987) Cell fate in the *Drosophila* ommatidium. *Dev. Biol.* **123**: 264–275.

sevenless

Basler, K., Yen, D., Tomlinson, A. and Hafen, E. (1990) Reprogramming cell fate in the developing *Drosophila* retina: transformation of R7 cells by ectopic expression of *rough*. *Genes Dev.* **4**: 728–739.

Basler, K., Christen, B. and Hafen, E. (1991) Ligand-independent activation of the sevenless receptor tyrosine kinase changes the fate of cells in the developing *Drosophila* eye. *Cell* **64**: 1069–1081.

Tomlinson, A., Bowtell, D.D.L., Hafen, E. and Rubin, G.M. (1987) Localization of the *sevenless* protein, a putative receptor for positional information, in the eye imaginal disc of Drosophila. *Cell* **51**: 143–150.

seven-up

Heberlein, U., Mlodzik, M. and Rubin, G.M. (1991) Cell-fate determination in the developing *Drosophila* eye: role of the *rough* gene. *Development.* **112**: 703–712.

Mlodzik, M., Hiromi, Y., Weber, U., Goodman, C.S. and Rubin, G.M. (1990) The Drosophila *seven-up* gene, a member of the steroid receptor gene superfamily, controls photoreceptor cell fates. *Cell* **60**: 211–224.

SOURCES OF FIGURES

For details, see above.

Figure 8.1 Photograph courtesy of R. Carthew and U. Heberlein.

Figure 8.2 After Ready (1989).
Figure 8.3 After Horridge and Meinertzhagen (1970).
Figure 8.4 After Lawrence and Green (1979).
Figure 8.5 After Baker and Rubin (1989).
Figure 8.6 See Ready (1989), Tomlinson (1988) and Tomlinson and Ready (1987).
Figure 8.7 After Tomlinson (1988).
Figure 8.8 From Tomlinson *et al.* (1987).
Figure 8.9 Photograph courtesy of A. Tomlinson.

Conclusions

'Gradually the scene grew clearer, and we could pick out individual objects. First, right opposite to us — we had been conscious of them all the while but refused to believe in them — were three great gilt couches, their sides carved in the form of monstrous animals ... with heads of startling realism.' [13]

Embryologists and geneticists never used to see eye to eye. As I have indicated in this book the two disciplines have now become united in a new subject formed by the fusion of developmental genetics with molecular biology. So what have been the main changes in perspective?

Some things have become more complicated. Setting up the antero-posterior gradient and interpreting it has, already, many more steps than previously envisaged. The special system to establish the head and tail was a complete surprise. The ways in which single genes are used and reused makes a nonsense of much traditional developmental genetics which thought of a gene working at one stage of development and having a critical period of function and a unique focus of action.

Moreover, we may have underestimated the amount of genetic information needed to design and organise rather than to build. Many of the genes in this book have been selected because their mutant phenotype hinted at a role in the design of the fly. Most of these have turned out to encode proteins whose sequence suggests they function by binding DNA. There are others whose sequence is not yet interpretable but because they are basic proteins, and because they are localised in nuclei, may also be engaged in the regulation of other genes. Regulatory genes like these appear to be particularly important in laying down the body plan of the young embryo, or in organising pattern formation in later stages. There are still others whose role appears to be in the transmission of signals between cells — signals involved in allocating cells to specific developmental fates. These signals are often received at the membrane and conveyed into the nucleus, again sometimes by special genetic systems. There are so many such genes that one begins to wonder what proportion of all are engaged in design and control.

There is another general point that relates to this argument. Take the eye as an example: about 30% of all lethal mutations, when examined as clones in the eye, damage eye development. If about 90% of all genes can mutate to give a lethal, as is generally thought, this means that nearly one-third of all genes make some contribution to eye development. Of course, some of these are probably housekeeping genes, such as the universal proteins, actin and tubulin. However there is still room for genes that are important in eye design but are not specific to the eye. An example is *Notch*. Yet at the moment most of the research on eye development is concerned with a small handful of genes, so much so that if a receptor is identified there is a tendency to look for the ligand in the same handful. This may prove to be wishful thinking; it may be necessary to clone and study hundreds rather than tens of genes before the full story of eye development is understood.

The very long regulatory regions of genes, the conservation of many bits of these (when different species are compared) and the multiplicity of binding sites shown up by footprint experiments, all suggest that the majority of genetic information is engaged in regulation. Perhaps no surprise if you think of the enormous logistical problems there must be; compare the making of a fly with the building of a moon rocket — those who actually construct the components are in a small minority. This view of genes and what some do is considerably removed from the one gene:one enzyme idea, which had such simplicity and persuasive force.

Some other elements of recent progress make the overall task seem less daunting. The most valuable researches have always opened more doors than they have closed. Molecular genetics is no exception, but one difference nowadays is how often the doors, when opened, reveal a room full of scientists beavering away on projects apparently unrelated to *Drosophila*. There are many examples, here is one:

In 1983, Wilcox and Brower set out to identify antigens whose distribution in development would not depend on cell type but correlate with domains of organisation. In the wing disc of *Drosophila* they found two such antigens, one expressed dorsally and one ventrally on the wing blade. Eventually, the antigens were purified and the genes identified and sequenced — they turned out to be integrins, a family of proteins linked to the cell surface (such as the fibronectin receptor of vertebrates) which are needed to attach cells to each other and to their substrates. In vertebrate systems, much more molecular work had been done on integrins, but no genes were identified or mutations known — a serious handicap. The results on *Drosophila* and vertebrates can now be brought together to improve our understanding of cell biology in both systems. And this is needed, for the cloning and sequencing of genes in *Drosophila* too often leads to a dead end, an end brought about because the sequence and the distribution of the gene product in embryos does not always clarify the mutant phenotype or the wildtype functions of the gene — often because of a lack of knowledge of cell biology.

Indeed, it is striking how many genes cloned in *Drosophila* have their homologues in animals as far away as vertebrates. This also works the other way round; for example, many of the oncogenes of vertebrates are homologous to genes in flies and sometimes the sequence identity is impressive. Moreover, some vertebrate genes, when transformed into flies, can do the jobs done by the homologous genes in the fly. These facts make a strong molecular argument for universality, not only for components but also for at least some mechanisms of design. If much is universal, everything becomes more relevant to everything else, and thus problems become more soluble.

These are stirring days in developmental genetics; the inkling of how the mother and the zygote combine to make the ground plan of 'the small gilded fly' [14] is just one piece of evidence that the new thinking and the new methods work. There **are** glimpses of clarity — enough to see the immensity and beauty of the problem and enough to know that there is still a long and challenging journey ahead.

History short stories

The segmentation genes

The very first *Drosophila* mutation was found in March 1910 by Thomas Hunt Morgan and was called *speck*. Two months later the famous *white*[1] allele turned up spontaneously in one of his bottles. There was no way of making mutations then; X-rays were first used in the 1930s and the much more effective chemical mutagens not until the 1960s. For this reason *Drosophila* geneticists worked mainly with viable and fertile alleles and for several decades their main preoccupation was the study of inheritance rather than development.

In Chapter 1 I explain how and why the newly laid *Drosophila* egg contains most 'ingredients' for making a larva, but, largely, lacks the 'recipes', products of genes needed to organise that development. This presents the scientist with a golden opportunity to identify and study those genes which are required for design. When a gene specifically required for spatial organisation is missing or defective, embryogenesis can still continue and the cuticle may be properly formed — only its **pattern** is awry. This allows pattern mutations to be selected. The principle was first used by Antonio Garcia-Bellido and colleagues, who searched chromosomes for homeotic genes (see p. 211) in this way.

Figure A1.1 Thomas Morgan. Photograph courtesy of the Bettmann Archive.

However the approach really took off when Christiane Nüsslein-Volhard and Eric Wieschaus began their search for segmentation genes in 1978. Some readers will know the impact of their work on the field, but the others should realise that it has been incomparable. They **systematically** searched the genome for mutations that altered the pattern of embryonic development. Since many of the mutations were lethal and autosomal they had to do an F2 screen (see Box 5.3, p. 131) — which means that a mutagenised chromosome (bearing a mutation, m^-) had to be carried and amplified in the second generation. Heterozygous m^-/m^+ males and females from the F1 generation were crossed and their offspring screened for dying embryos which produced cuticle with intriguing defects. Many thousand flies with mutagenised chromosomes were screened and, amongst that population, a typical gene was hit several times — showing that the majority of genes that could be found in this way were identified. It was a mammoth effort involving a great deal of organisation. Not only did each lethal phenotype have to be spotted down the microscope, it was important to concentrate attention on interesting phenotypes (a lot of intuition here), to map the new loci and to classify the genes in a conceptually useful way. In 1980, Nüsslein-Volhard and Wieschaus published the outline of their results. Amongst other things, they defined the gap, pair rule and segment polarity classes of genes.

Later, in collaboration with Trudi Schüpbach and colleagues, Nüsslein-Volhard and Wieschaus extended their search and systematically looked for maternal-effect lethals that did interesting things to the cuticle of larvae. Remember, maternal-effect lethals are mutants whose phenotype is determined largely or entirely by the genotype of the mother — they cannot be rescued by wildtype sperm (Chapter 1). Their figures, based on large numbers of mutagenised chromosomes and the several alleles they produced at each locus, give an estimate of only about 40 for the total number of genes in this class. This fact

Figure A1.2 Janni Nüsslein-Volhard and Eric Wieschaus, 1990.

argues strongly that embryos are elaborated from relatively simple beginnings; a few maternal-effect genes, a large number of zygotic genes. The alternative traditional view, that the egg is a highly complex structure packed with many determinants, would have needed many more maternal-effect genes than the 40 found. Amongst the mutations were a class of phenotypes called coordinate mutations, because they affected either the anteroposterior axis or the dorsoventral. Examples are *bicoid* (p. 28) and *Toll* (p. 44).

Every 2 years the main international meeting on *Drosophila* developmental genetics is held, happily, in Kolymbari, Crete. In 1980 Nüsslein-Volhard and Wieschaus announced their findings there and in the last five of these meetings the proportion of talks dealing with the molecular cloning and other aspects of 'their' genes has increased steadily. In 1990 it had reached about 50%.

The morphogenetic gradient

One of the most long-lived and illuminating stories in the whole of developmental biology is that of the morphogenetic gradient. A simple and powerful idea, it arose directly out of Thomas Hunt Morgan's experiments on regeneration in an annelid worm. At the end of the last century, Morgan decapitated a lot of worms; he found that the speed with which they made new heads depended on where the cut was made — 'neck' cells made a new head quickly, body cells more slowly. He concluded that there were factors that varied **quantitatively** along the main axis of the body.

'I think . . . we are dealing here with something that is connected with the organization of the worm itself. Perhaps for want of a better expression, we might speak of the cells of the worm as containing a sort of stuff that is more or less abundant in different parts of the body. The head stuff would gradually diminish as we pass posteriorly, and the tail stuff increase in the same direction. We should also think of this stuff in the cells as becoming active during regeneration. Where there is much of the head stuff, the cells can start sooner to regenerate anteriorly: where there is less it must increase first to a certain amount or strength before the part can begin to regenerate. I do not pretend that this explains anything at all, but the statement covers the results as they stand.' [15]

Note that he thought this behaviour revealed 'the organisation of the worm itself' and this idea was developed in the next 70 years with the help of many experimental animals such as protozoa, hydra, planaria, sea urchins and frog embryos. The essential model of a gradient is that some factor, possibly (but not necessarily) a chemical substance or 'morphogen', varies in a monotonic way along the length of the body. The local level or scalar of this gradient (concentration in the case of a substance) determines the local response. So the scalar of the gradient may tell the cells where they are in the body and therefore determine the pattern of development, or it may determine the capacity to regenerate.

The number of embryologists working on pattern formation has always been small and they have tended to be eccentric people. Thus odd ideas about gradients were frequent. Charles Child wrote an extremely fat book on gradients to advertise his idea that gradients were essentially metabolic — meaning that they involved ordinary metabolism rather than special molecules and specific mechanisms. One of the experiments presented was with some poor flatworms. These animals were placed in solutions of potassium cyanide and it was noted that they died from the head backwards — evidence that the metabolic rate was higher at the front of the body and literally decayed away towards the tail. Another idea, which still has its supporters, is due to Meryl Rose who studied the hydroids that have amazing properties of regeneration — even tiny pieces of body can remake heads and tails and then go on to make a colony again. The problem addressed by Rose is that, if the cells in every piece of the body can make heads and tails what stops them from doing so when they are *in situ*? He suggested that the body might be made up of several discrete parts (such as head, neck and trunk) and that each part produces a substance that inhibits other parts from differentiating in the same way. Thus the head produces a diffusible head inhibitor and, so long as the head is present, other parts cannot become head. Remove the head and all parts will be released from inhibition, and all will try and transform towards head. The model needs the idea of hierarchy, an intrinsic order of potential in the body axis. So neck, being next to head in the body, is also the next in the hierarchy to head; when inhibitor is removed it will reach the status of head first. The model is not very pleasing because there is no help in estimating whether a hydroid divides itself up into two parts or 2000, with two or 2000 inhibitors.

Some of the best analytical experiments on gradients were done on insects, particularly by Klaus Sander as a graduate student in the late 1950s, and these are described in Chapter 6. Growing up in the German school of 'Developmental Mechanics' Sander found his gradient models ill-received. His supervisor, Gerhard Krause, was a staunch supporter of Friedrich Seidel's theories of insect development, which did not feature gradients. When Krause heard about Sander's gradient theories he said 'I thought you were supposed to prove that Seidel's ideas apply to the leaf-hopper, too'. It is a good thing that most graduate students do not make discoveries under contract. Sander also remembers a fellow student drawing a cartoon of a little dog biting into a large slanting leg — representing Sander himself and his gradient model. As Sander explains: 'it was not only little dogs that disliked gradients'.

Earlier, one of the most influential big dogs had been Hans Spemann; Sander has pointed out to me that Spemann, in his famous 1936 book *Experimentelle Beiträge zu einer Theorie der Entwicklung* (Springer, Berlin), explained his anathema to gradients: 'Now to me it appears that a gradient can only exert an effect if something is really "flowing"; for instance a stream of water or an electrical current. Even the steepest hillside cannot, as such, drive a water mill.' I now quote Sander: 'Spemann's argument was elegantly refuted by

Figure A1.3 Klaus Sander when a student at Tübingen University, 1958. Photograph courtesy of K. Sander.

Leopold von Ubisch, the first person who, as far as I know, explicitly discussed the relation between gradients and genes. Referring to Spemann's metaphor of a mountain he argued that each altitude has its own flora, presumably because the slope of the mountain cuts across a gradient of temperature. The climatic gradient on the mountain selects, at each level, a few plants from the many. Going further, he pointed out that in the embryo at each level, the cytoplasmic gradient could activate a subset of the entire genome; there would be no need for anything to flow.' von Ubisch's suggestion that gradients can pattern gene expression provides the theme of the work discussed in Chapters 2 and 3.

It was also in insects that the relationship between cell polarity and gradients was first appreciated. As I discuss in Chapter 6 the idea is that the **vector** of the gradient (the direction of slope at any point) determines the local cell polarity which is expressed, for example, in the orientation of bristles. The experiments which led to this were done by Hans Piepho, Michael Locke, Hildegard Stumpf and myself. Locke and I were both students of **the** insect physiologist, V.B. Wigglesworth, and we followed up on his experiments described in two outstandingly original papers on pattern formation in the bloodsucking bug *Rhodnius* (1940). Stumpf came from the school of Piepho whose interest in cell polarity began while he was sunbathing on the beach and idly began to wonder why the hairs on his upper and lower arm pointed in opposing directions. Biologists should take holidays. Stumpf and I came to the same conclusions about concentration gradients, vectors and cuticle polarity independently, although her grasp of the problem was more complete than mine. In 1967, as she began some of her most probing experiments, she was killed in a car accident.

With the rise and tremendous success of molecular biology in the period from the 1950s to the 1970s and the appearance of this new and rather intransigent species of modern scientist, the relatively fuzzy gradient models and model-builders fell into disrepute. Lewis Wolpert, who had worked on gradients in *Hydra*, recalls proposing a general role for gradients in pattern formation at a meeting in Woods Hole in 1968. Afterwards he found himself ostracised by the 'very hostile' response to the subject of his talk. The problem with gradients was we did not know what the gradient substance or 'morphogen' was and this made the models too abstract for current taste. It was not the only problem, though — botanists had already identified morphogens in plants, such as auxins that pattern the development of new shoots, but this did not make gradients palatable. The idea was just too imprecise and too simple, to be accepted. 'Gradient' was a dirty word in the 1960s and 1970s.

The successful identification of the first definitive morphogen (the *bicoid* gene product) did not happen until 20 years after Wolpert's talk and these experiments are described in Chapter 2. As I write, gradients are respectable again, people are now happy to resort to them on the least provocation. However, we still do not know how gradients work in true multicellular systems, although there are some promising leads.

engrailed: the life history of a gene

'Fools change in England,
    and new fools arise;
For, tho' th' immortal
    species never dies,
Yet ev'ry year new maggots
    make new flies ...' [16]

The history of the *engrailed* gene is instructive, for it illustrates the predominant role played by fashion in science, not just yesterday but today, too. What is considered crucial and exciting by one generation is soon deemed to be irrelevant and boring by the next. And scientific generations seem to be shortening.

The first allele of *engrailed* was spontaneous and it turned up in Reidar Eker's laboratory in Norway in 1926. We now know that spontaneous alleles are often caused by mobile elements and, indeed, *engrailed¹* has proven to be a 7 kilobase insert in the 3' regulatory DNA. What was important at the time was inheritance; mutants were valued according to their usefulness in crosses designed to follow chromosomes or parts of chromosomes through the generations. Thus Eker concluded 'The *engrailed* flies are of good viability and fertility. They can be separated easily and accurately from wildtype flies. *engrailed* should accordingly be a useful mutation for general work.' [17] *engrailed* is a heraldic term and the mutation was named after the cleft or notch in the scutellum. Probably because one clear phenotypic difference is sufficient for inheritance studies, other aspects of the phenotype, such as the duplicated sex comb in the T1 leg, were illustrated but not commented on. At that time, many *Drosophila* geneticists behaved like collectors; new mutations tended to be written down, written up and then cherished for the future.

During the period 1930–1950 *Drosophila* genetics began to move a little towards development. Still the emphasis was often on the mutant (rather than the wildtype) in spite of George Beadle and Boris Ephrussi's beautiful experiments on *Drosophila* which suggested that a gene represented a discrete molecular function: the beginnings of the one gene, one enzyme idea. The standard approach of developmental genetics at that time was to make a detailed description of the mutant phenotype under different conditions and this was sometimes done without sparing a thought for the wildtype function of the gene. Adair Brasted (1941) chose the gene because 'the variability associated with the phenotypic effects of *engrailed* could be investigated experimentally, since each is sensitive to environmental changes.' [18] This thought is followed by 26 taxing pages on the effect of body weight, temperature, crowding and intersexuality on the number of sex comb teeth — he was the first to report that *engrailed¹* affects the T1 leg and gives an extra row of teeth.

Twenty years later, influenced by the ideas of her colleague and mentor Curt Stern, Chiyoko Tokunaga looked at the cell autonomy of *engrailed¹*. In *engrailed¹* mutant flies, there are extra 'secondary' sex comb teeth and these form in what we now know to be the posterior compartments of the T1 legs. She made clones of cells that were homozygous for *engrailed¹* in a wildtype background. She found that if a small *engrailed¹* clone appeared in the place where the secondary sex comb developed then that clone developed a few teeth, in other words the *engrailed* mutant pattern developed cell-autonomously.

Tokunaga also remarked 'cell lineage studies in mosaics suggest an early developmental separation in the leg discs of the regions determining primary and secondary sex combs.' [19] This is tantalising because she might have discovered developmental compartments; she noticed the lineage segregation but did not follow it to its logical conclusion. Another scientist, Peter Bryant, also nearly discovered the division of the wing into an anterior and a posterior compartment. He had a fair number of clones in the wing, none of which transgressed what we now know to be the compartment boundary (some examples are shown in Figure 4.2). Even so, it would have been very difficult to spot the compartment boundary with the methods he used; the *Minute* technique (p. 83), which gave much larger clones, had not been invented and the smaller clones he had reveal the compartment boundary only to an observer with the advantage of hindsight. Also he was 'only ready to find boundaries that corresponded with morphological subdivisions. The dorsal/ventral one was clear of course'. These near misses by Tokunaga and Bryant are instructive because they can be related to the general scepticism that greeted the paper, published in 1973 by Antonio Garcia-Bellido, Pedro Ripoll and Gines Morata, that used the *Minute* technique and first described the anteroposterior compartments in the wing. Very few believed they were right. It seems that in a field like developmental biology, where we know little and, perhaps, because we understand less, expectation and prejudice play far too big a part in judgement.

In 1967 Garcia-Bellido went as a postdoc to Ed Lewis' laboratory and there he started two practices, one technical and the other intellectual. To appreciate the technical innovation we need a little background: scientists are very conservative, they tend to stick to the techniques and equipment they know. Because large numbers of living flies have to be sorted, the geneticist's standard aid was the dissecting rather than the compound microscope. One consequence of this preference was the image of eye development, it remained incorrect for 50 years because, from the 1920s until the early 1970s, those who made mosaic eyes never cut a section through one and looked at it under the compound microscope. Had they done so they might have found the ommatidia were not clones, as had been assumed. So, you will understand that when Garcia-Bellido took to describing genetic mutations with the compound microscope, it was an original and important move; important because the description of mutants

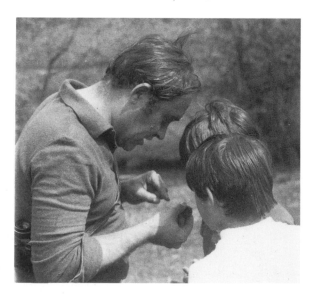

Figure A1.4 Antonio Garcia-Bellido with two of his sons, 1974.

and mosaics was brought down to the cellular scale. One of his discoveries was that the mutants *forked*[36a] and *multiple wing hairs* are excellent cell markers in the wing and could be used in experiments on cell lineage. Intellectually, he and Lewis were influential in shifting the perceived aim of developmental genetics from describing mutants to determining the wildtype function of genes.

To return to the *engrailed* gene; Lewis, in a conversation with Garcia-Bellido, pointed out that *engrailed*[1] affects the wing and seems to transform posterior parts towards anterior pattern. In 1972, Garcia-Bellido and Pedro Santamaria reported a mitotic recombination study of *engrailed*[1]; they concluded: 'The mutant *engrailed* causes the overall transformation of posterior structures and pattern organisation into those typical of anterior regions. In this sense *engrailed* represents a homeotic mutant.' [20]

In 1974, Morata joined me in Cambridge and we conducted yet another analysis on *engrailed*[1] using two improvements: one was the *Minute* technique which produced exceptionally large clones and the other was a new marker mutation called *pawn* which labelled every single cell on the wing surface. We also knew that wings were divided into anterior and posterior compartments and therefore wished to test whether the *engrailed*[1] mutation affected only posterior cells, which might imply that *engrailed*[+] was a selector gene (see Chapter 5) active in the posterior cells to distinguish them from anterior ones. Opinions in the laboratory canteen were split and a bet was set up. Some thought that *engrailed* might only pattern a preexisting matrix, that it might simply 'colour-in' the posterior compartments, without having a role in the establishment or maintenance of the compartments themselves. Others thought that it might also be important in defining the compartment boundaries. As it turned out, the results showed that *engrailed* is a selector gene necessary for all the posterior cells, not only in determining the pattern they construct, but also in making them distinct from the anterior cells — a distinction that is crucial for the maintenance of the boundary between anterior and posterior compartments. As a result of these findings the *engrailed* gene became the cornerstone of the compartment hypothesis (see Chapter 4), at least for the believers.

There were alternative viewpoints: some thought that the anterior and posterior cells of the compartments did not arise as neighbouring groups, but instead came from more remote cells that migrated together in some genetically uninteresting way — the 'construction hypothesis'. Others imagined that the *engrailed* mutant phenotype was due, not to the failure of a process that identified posterior cells, but to cell death and regeneration — events associated with duplication of patterns in imaginal discs.

The *engrailed*[1] allele had been around for 50 years, but no other alleles were known. In 1975 Thomas Kornberg joined Morata and me and he began to make more alleles; over the years he found many and most of them were lethal. However, the lethal alleles behaved, in mitotic recombination clones, in a manner concordant with *engrailed*[1]; they affected posterior but not anterior cells. When Christiane Nüsslein-Volhard and Eric Wieschaus did their screen (p. 202) they found more *engrailed* lethal alleles and, on the basis of the rather messy embryonic phenotype, classified *engrailed* as a pair rule gene (p. 69).

It is salutary to think that the slowness of the progress over the first 50 years was not due to the lack of technical knowledge. An ingenious scientist could have done all the experiments in 1929 — mitotic recombination experiments had already been invented then although they were not understood until Curt Stern's paper of 1936. However the next step, cloning the gene, did depend on technical advances. Kornberg's group had been 'walking' from a start point recognised on the polytene chromosomes by *in situ* hybridisation with a tRNA probe. Independently, Anders Fjose and Walter Gehring had fished out *engrailed* as a homeobox-containing gene and, in spite of all the preceding 60 years, the two papers describing the cloning, sequence and pattern of expression of the gene came out simultaneously. As expected from the mitotic recombination studies, *engrailed* was expressed in a subset of the segment in the embryo and in the posterior half of the disc. The subset of cells in the segment did turn out to be posterior, but this became clear only after the segment/parasegment confusion had been resolved (p. 91). The genetic mosaic experiments had led to the expectation that, in the embryo, there would be anterior and posterior cells and that the latter would express and depend on the *engrailed* gene for their identity. This was supported by antibodies against the *engrailed* protein which reveal a pattern of stripes that first appear in the young embryo and persist indefinitely.

In the last few years we have made rather little progress in answering the many outstanding questions about *engrailed*. For example, Kornberg has discovered that *engrailed* has a sister gene (*invected*) which is adjacent to *engrailed* and has considerable sequence homology with *engrailed*, including a homeobox. Its pattern of expression is also similar, but mutations specifically in *invected* are unknown. Would clones of cells that lack both *invected* and *engrailed* be completely transformed to the anterior state? Or is *invected* responsible for a different set of posterior compartments (in the head?)? Are there other completely different genes which, together with *engrailed*, specify the posterior state? Does the anterior state depend on selector genes of its own? Is segmentation in other animals dependent on *engrailed*-like genes establishing subsets of posterior cells?

There are major questions about all the selector genes, including *engrailed*. How do they produce the changes in pattern for which they are responsible? What is their role in cell affinities? How are they regulated and how do the different proteins interact with the DNA, with other genes, and with each other? In another 50 years, will our present interest in *engrailed* look like an aberration; will *engrailed* be forgotten or will there be another aspect of it that seems all-important to a new generation of scientists?

The history of the bithorax complex

In 1915 Calvin Bridges, one of the first fly workers to join Thomas Hunt Morgan in the fabled fly room, found *bithorax[1]*, a mutation that transformed part of the haltere towards wing (see Figure 5.1). This mutation was called 'homoeotic' — William Bateson's word for a malformation that substituted the pattern of one region; 'not that there has merely been a change, but that something has been changed into the likeness of something else.' [21] He was thinking particularly of systems of reiterated elements, such as segments in the arthropod, petals and sepals in a flower or teeth in a vertebrate. Homeotic mutations are encouraging, because they raise the clarifying prospect of a class of controlling genes responsible for large chunks of the body pattern. They also impressed because the mutations produce massive anatomical transformations; it was even thought such mutations could allow the sudden generation of new animal groups during evolution — an idea that looks increasingly implausible (individuals produced by such mutations are very unfit!). Nevertheless *bithorax[1]* was filed away and little progress made until Ed Lewis started work on the locus in 1946. At a time when most geneticists were busily studying the details of inheritance, recombination and chromosome mechanics with rigorous quantitative methods, Lewis, a real pioneer, began a lifetime's tussle with deeper problems that were not ready for such a disciplined, and therefore circumscribed, approach. These problems were both genetic and developmental

Figure A1.5 Pam and Ed Lewis, c. 1980. Photograph courtesy of E.B. Lewis.

and the genetic conundrums he faced made it difficult to decipher the wildtype structure of the gene.

In the *Drosophila* chromosomes, there are a large number of 'doublets', that is pairs of neighbouring chromosome bands that look very similar. Bridges had suggested that each arose by the duplication of a gene, this being one of the main ways genes could increase and diversify. Lewis' PhD thesis had demonstrated the correctness of this with respect to the *Star* gene and, after a spell of war duty spent weather forecasting in Okinawa, Lewis returned to the problem, selecting two more sets of mutations for study. One was the *bithorax* locus (see Figure 5.4). The *bithorax* mutations were unusual, in that there were many dominant alleles that carried complete or partial homeotic transformations and, above all, the complementation pattern was puzzling. Some mutations suggested the whole system of bithorax genes was integrated, while others pointed to separate and separable elements of function. Lewis realised that both were true, that there are different genetic elements but they are interlocked and function as a complex. Some molecular biologists did not agree: 'It was hard persuading people we were dealing with gene complexes. They told me I was simply dealing with missense and nonsense mutants **within** a protein and that all we were doing was mapping sites within a single protein coding unit!'

Nowadays Lewis might have lost his grant, but those were more permissive times; he was undeterred and faced up to the daunting task of understanding the bithorax system of genes. He spent years building up a large collection of mutations. For analysis, he relied largely on viable alleles which meant, as we know now, that he was working with weak partial mutations that affected control regions. At the time, not knowing this, he followed the line of thinking that each different class of mutation caused the loss of a discrete wildtype function, even a wildtype substance. As the range of mutant phenotypes was extended, so his models inevitably became more complex and, naturally, tended to follow the fashion of the time. For example, in the early 1960s, the work on the operon led Lewis to speculate that Ubx^- might be an operator-null mutation with the *bithorax*$^+$ and *postbithorax*$^+$ functions (see p. 108) as part of the operon. Thus, *Contrabithorax*, a gain-of-function mutation resembling *Haltere mimic* (see Figure 5.1), was thought to be an operator-constitutive mutation.

In the 1950s and 1960s Lewis realised that the wildtype role of the *bithorax* genes was in specifying 'developmental pathways' — the routes followed in building the characteristic pattern of the different parts of the body — and that these genes worked **locally** to regulate this. This was a crucial leap in understanding. He made some genetic mosaics with *Ultrabithorax* (p. 108) and showed that the wildtype gene was required in the T3 segment that makes the haltere. He made clones of cells that carried a *bithorax* mutation and a genetic marker; these suggested that the wildtype genes are required autonomously, meaning that only mutant cells are transformed and none are rescued by nearby wildtype cells. The marker used identified only bristles and not epidermal cells so he could not be absolutely sure of this. Much later, in 1976, Gines Morata and Antonio Garcia-Bellido used a better marker and proved that this local requirement actually means that the bithorax complex of genes work within specific groups of cells and that the differentiation of each cell depends on which gene is expressed in it. In other words, the requirement for the genes is 'cell autonomous'.

Lewis noted that the *bithorax* mutations mapped in an order that corresponded with the order of the parts in the body, the leftmost mutations

Figure A1.6 Gines Morata, 1980. Photograph courtesy of G. Morata.

affecting the most anterior body parts (p. 116). This correspondence between the order of the genes and the order of the body parts has now attained almost mystical status. It is true of the vertebrate homologues of the bithorax complex and very few papers fail to remind the readers of it. The conservation of gene order in so many groups cannot be without significance. Surely? And yet, in *Drosophila*, genetic experiments show that at least two of the genes, *Ubx* and *Abd-B*, can each work well when displaced and alone. If they can work independently, it is not clear why they have remained together — especially on the choppy seas of evolution. Indeed, Lewis believes that the genes **are** subtly interdependent and that the correspondence between the order of body parts and sites of mutations affecting them is no accident: 'one cannot sweep under the rug the striking colinearity that extends to the vast *cis* regulatory regions. Not only do the latter constitute some 95% of the bithorax complex against some 5% of the transcribed DNA, but they tell the homeobox-containing genes what to do'.

It was not for many years that Lewis looked at the effect of lethal mutations on the cuticle phenotype of the moribund larvae. This was an imaginative and important move. Now he saw that the bithorax complex (or BX-C as he then called the gene system) had a much wider role than previously thought. Using a scanning electron microscope he saw the effect of deleting the entire BX-C. He found that many body segments are transformed and, looking at deletions, proposed that there are a large number of discrete genetic elements each being responsible for a segment, and even sub-elements responsible for particular parts of the pattern (denticle belts, Keilin's organs, etc.). The results were in logical accord with his experiments on the *Ubx* gene. In *Ubx*⁻ larvae, he knew that T3 and A1 were transformed to T2, now he found that, in the absence of the entire BX-C, all the segments of the abdomen became thoracic. There had to be more genes. By pure genetic experiments he showed that the leftmost genes have the longest range, are activated most anteriorly and affect the pattern all the way posteriorly; mid genes are activated in the middle but still affect parts back to the posterior end. This means that each part of the body is specified by a different set of active bithorax genes and, therefore, that there is a **combinatorial** code that defines developmental pathways.

The idea that homeotic selector genes (Chapter 5) act together in combination to specify pattern has prospered in recent years. Most notably it has been taken up by Gary Struhl, who showed how helpful the idea is in understanding local mutant phenotypes. He also showed, by experiment, that unusual combinations which do not exist anywhere in normal flies lead to 'nonsense code words' and messy, abnormal patterns. Struhl argued that the BX-C genes do not act by specifying individual elements in the pattern one-by-one, instead they work on groups of cells as a whole to influence their pathway of development.

However, there were two types of experiment that even Lewis did not do. He did not do the detailed study of wildtype cell lineage that would have told him that the 'developmental pathways' affected by BX-C mutations were in particular sets of clones, in compartments (Chapter 4). Nor did he try a conventional mutagenesis on the BX-C, collecting lethal alleles. This key experiment was taken up by Ernesto Sanchez-Herrero and Morata and the results changed and simplified the picture of the BX-C. They took flies carrying chromosomes in which the entire BX-C and some adjacent genes are deleted and crossed them to flies carrying chromosomes subjected to random mutagenesis. Occasionally, the combination of the two chromosomes was lethal, implying a new mutation in the region of the deletion. Many of these new mutations were collected and mapped and their lethal phenotype in embryos and their viable phenotype in clones of cells in the adult were studied. It became clear that there are only three independent genes in the BX-C and that many of the numerous genetic elements inferred by Lewis were due to lesions in the regulatory regions of these few genes. Morata's paper was too simplifying to convince immediately, but it was made more palatable by the discovery at the same time that there are just three homeoboxes in the BX-C, that is one for each of the three genes *Ubx*, *abd-A* and *Abd-B* (Chapter 5).

Morata's view is now widely adopted, as is the acceptance that the three BX-C genes are generally expressed and required in sets of **parasegments** rather than segments. The first sign of this came when Morata and Steven Kerridge made clones of adult leg cells that were homozygous for Ubx^-. These clones, when made early in development, affected T2p, T3a, T3p and A1a, transforming them to T1p, T2a, T1p and T2a respectively. The effect of Ubx^- clones on T2p came as a shock as it had been accepted that Ubx^+ was responsible for the T3 segment and had no role in the T2 segment. I was visiting the Morata laboratory at the time and I remember looking down the microscope and greeting the result with incredulity and hilarity. I took the view that the BX-C, which was already impossibly complex, was in danger of becoming unintelligible and advised Morata to quit the field before he was overwhelmed. Even Lewis was puzzled and, on December 12, 1980, he wrote to Morata 'I am at a loss to explain your very interesting results'.

The next step forward came when Struhl looked at flies in which the BX-C had been split into two pieces, each piece being on a different chromosome. The flies developed normally, showing that the BX-C did not need to be intact to work. He studied embryos that had only the leftmost part of the complex and noticed that the embryos formed a chain of compartments T3p + A1a, T3p + A1a and so on, and suggested that the 'BX-C genes may normally act on segmental units extending from the middle of one segment to the middle of the next.' [22] A similar conclusion was drawn by Rob Denell and colleagues. Using several criteria, Alfonso Martinez-Arias and I then proposed that these segmental units were not only the modules of BX-C function but were actually the archetypal segmental unit, that they were formed earlier, and had a more fundamental role in building the fly, than segments. We called them para-segments. Once it had been fully appreciated that all three BX-C genes reacted to a parasegmental register, most of the confusion receded. This generalisation has been amply confirmed by experiments with antibodies against the proteins encoded by *Ubx*, *abd-A* and *Abd-B* — although there are minor complications (see Chapter 5).

Cloning the BX-C was a major achievement, particularly as it was the first *Drosophila* gene cloned 'from the chromosome' — without any prior knowledge of the product. In order to clone this forbiddingly large stretch of DNA, David Hogness' group had to develop the method of 'walking' along the chromosome. The cloning illuminated the genetics, for the regions that coded protein were identified, introns mapped and the large *cis* regulatory regions recognised. It then became possible to classify the mutations in molecular terms — do they damage coding or regulatory regions, are they deletions or transpositions? The result is that the bewildering variety of BX-C mutations now make more sense. Cloning is also an essential prerequisite to making probes and antibodies for the gene products and plotting their distribution in time and space.

The discovery of the homeobox

I am reliably informed that genetics and developmental biology are more fun than experiments with molecular biology. While it is true that the information gained from molecular biology can be very hard it is also the result of very hard work. In spite of the technical advantages of *Drosophila*, cloning, identifying and sequencing a gene can easily take up all of a PhD student's 3 or 4 years. The advantage of molecular biology is that the information can sometimes be indisputable and sound — an example would be the sequence of a protein (provided it had been done correctly!). However, rock solid data show up how shallow our understanding of the subject is — consider how rarely a sequence can be simply translated into an appreciation of exactly what the protein does. Usually sequences, northerns and *in situ* hybridisations have to be interpreted against the background of soft information on protein function, cell biology and anatomy, as well as the numerous theories of pattern formation and morphogenesis.

An exception is the homeobox, a sequence of 60 amino acids that picks out a protein as being a DNA binding protein and the gene as controlling other genes. In all cases so far analysed it appears that homeobox genes are responsible for determining cell identity, in other words they are concerned with design, with the developmental pathways followed by cells. The homeobox signature, which is identified by nucleic acid hybridisation on filters, and must be confirmed by sequencing the candidate genes, has turned up in most multicellular organisms.

The homeobox was discovered in two laboratories independently. At about the same time (late 1982) each had cloned regions of the *Antennapedia* gene (p. 128). Matthew Scott, who was at that time in Bloomington, Indiana, in the laboratories of Thomas Kaufman and Barry Polisky, found that *Antennapedia* cDNA clones hybridised to *ftz*, which Amy Wiener had identified and located on their *Antennapedia* walk. Possibly influenced by Ed Lewis' ideas, Scott had requested cDNA clones from the *Ubx* gene from Welcome Bender and these also hybridised with the cDNA clones that linked *ftz* and *Antennapedia*. Cross hybridisations in themselves may not be very meaningful; it was known that there are many species of DNA sequence that are repeated hundreds of times in the genome. Therefore, Scott's next step was to pare down the hybridising piece, to sequence the DNA and then see if it represented a conserved stretch of amino acids. Scott was working late at night: 'My first sequences were not very good, I scribbled the sequence I could read onto pieces of paper, translating them in all frames. I colour coded the different types of amino acids to make it easier. The most exciting moment of my time in science came when I found a fragment of *ftz* sequence that matched a fragment of *Antennapedia* sequence'. He compared all three sequences, 'fixed all the frameshifts and worked out the entire 60 amino acid region. Each sequence corrected the other and it was obvious that most of the changes were in the third positions of codons.' This not only showed that the genes had a common conserved and therefore functional domain in their protein products, but proved that there were protein products. At that time the uniqueness of homeotic genes had led many to suspect that they might act as RNA, not protein. It seems absurd now, but homeotic genes had only just been cloned and there was a great deal of genetic evidence that these genes were very special.

In Basel, Switzerland, in the laboratory of Walter Gehring, Rick Garber and Atsushi Kuroiwa had isolated cDNAs from *Antennapedia*; Bill McGinnis and Michael Levine used these to see if there were portions conserved between *Drosophila melanogaster* and *D. hydei*. There were. Although they feared that the hybridisation meant little, they cut up the cDNA into smaller probes and although much of the hybridisation was to a highly repeated sequence (the so-called M or opa repeat), they found a small piece of 3' DNA that only hybridised to a few other fragments in the genome. In the spring of 1983 McGinnis and Levine then joined up with Ernst Hafen and, in Levine's words, 'Billy (McGinnis), Ernie (Hafen) and I feverishly screened genomic DNA libraries. We formed a human chain, with Billy mixing phage stocks and bacteria, Ernie adding the overlay and me pouring the concoction onto agarose plates. I frequently missed and hence the legend of our cloning the *Abd-B* gene from phage plaques grown directly on the bench was born (and is true)'. Clones isolated in this way were labelled and hybridised to polytene chromosomes *in situ* and to frozen sections of embryos. Some hybridised to near the BX-C and ANT-C and the embryo sections developed in the summer of 1983 told Hafen and Levine that they had isolated two more genes from the BX-C and *Deformed* from the ANT-C. As McGinnis remembers, 'By then we were sure that the sequence was not randomly distributed in the genome, and it acquired the homeobox tag.'

Meanwhile in the USA, Allen Laughon and Scott noticed that the sequence shared features with DNA binding proteins from bacteriophage, bacteria and yeast, and pointed out that it could fold into a structure that could embrace the DNA and might recognise the sequences of bases. A clever insight, which is only now being proved correct.

Figure A1.7 Matthew Scott, 1981. Left to right: Ernst Hafen, Mike Levine and Bill McGinnis, 1982. Photographs courtesy of M. Scott and E. Hafen.

Back in Switzerland, McGinnis and Levine were looking further afield: 'Mike suggested we might as well test some worms, so my wife Nadine walked to the local fisherman's bait shop and got a variety of insects and segmented worms. I had the DNA isolated even before we identified what species they were (Nadine kept a few of them around feeding them on rösti, pommes frites and Lifesavers candy). I put in some vertebrate DNAs (calf thymus, *Xenopus*, human) hoping for some cross hybridisation, but really thinking of them as negative controls. When I pulled the first blot out, there were obviously some strongly hybridising fragments in the human, frog and calf lanes. I was so excited my hands were shaking, but most were sceptical about the result and I was a bit disappointed.' However, McGinnis repeated the experiment successfully several times and people shelved their prejudices that invertebrates and vertebrates had to be completely different.

The discovery of the homeobox was greeted with great excitement. Universalists liked it, because the conservation across the animal kingdom suggested a commonality of mechanism. Geneticists liked it because, all along, the master homeotic genes had been thought to control other genes and here was concrete evidence that they might bind to DNA. For a time it seemed that homeoboxes were present only in animals that were segmented, and this, coupled with the homeotic (segment switching) characteristics of the Antennapedia and bithorax complexes, spawned the idea that homeoboxes were concerned with segmental identity. Since then, homeoboxes have been found more widely and often turn up in genes concerned with determining the state of cells, and some of these, such as the *rough* gene, are active in non-segmental systems such as the eye.

The homeobox has been particularly valuable as a guiding light into the vertebrate genome, and helped pick out a class of developmentally important genes there. In vertebrates there have turned out to be several clusters of homeobox-containing genes and some of these are strikingly homologous to their insect counterparts such as the bithorax complex. Not only are the sequences of the different elements in the vertebrate and insect gene clusters comparable, but the order of those elements on the chromosome is conserved. Moreover, the position of the anterior margin of expression in the body axis (*Ubx*, parasegment 5; *abd-A*, parasegment 7; *Abd-B*, parasegment 10) and the order on the chromosome correlate and this is also found in the vertebrates. In chicks the hindbrain is divided up into lineage units — called rhombomeres — that share some features with insect parasegments. The borders of expression of these homeobox genes in the vertebrates that show homology with the Antennapedia and bithorax complexes are also conserved in order, so that the most 3' genes in both insects and vertebrates are expressed in the most anterior segments, the most 5' in the most posterior segments. This remarkable conservation suggests that there may have been an ancestor common to flies and humans and that the body plan of that ancestor survives in the hindbrain of humans (the more anterior and posterior bits probably having been added on later) and the parasegments of insects.

Transforming flies

For some people, genetic engineering is a threatening idea: there is an image of evil men in white coats manipulating genes and producing a race of compliant warriors. The facts do not allow realisation of such an image, not even in the foreseeable future. However, a limited form of genetic engineering is possible in *Drosophila* and has already been extraordinarily useful; it depends above all on one particular discovery made by Gerry Rubin and Allan Spradling. First I describe the method: the aim of Rubin's and Spradling's experiments was to obtain flies in which a chosen piece of DNA is inserted into the chromosomal DNA. To do this you need a transposon or vector to carry the DNA. The one selected is known as a P element. The P element is derived from flies in the wild and has the capacity to hop about in the genome and there it encodes the necessary enzyme, a 'transposase'. Functional transposase is only produced in the germ cells, and so if it is to hop into a chromosome it will do so only there. The vector is modified; it carries a wildtype marker gene so it can be followed and has restriction enzyme sites so that the chosen DNA can be inserted. It does not carry its own transposase, which is provided on separate 'helper' DNA named 'wings clipped' — a nice way of indicating it cannot jump itself about.

DNA of both the transposon and the helper is mixed and injected into the posterior end of young eggs before the germ cells form. The helper DNA is

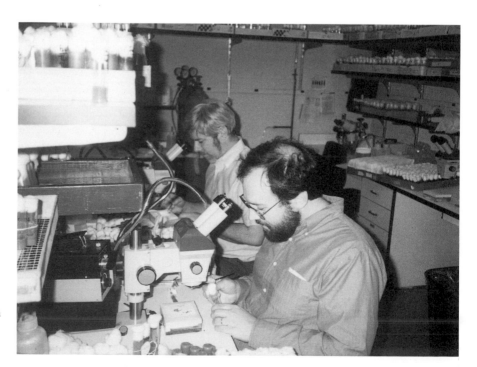

Figure A1.8 Allan Spradling and Gerry Rubin (nearest to camera), 1990. Photograph courtesy of G. Rubin.

transcribed and translated to make transposase which returns to the nucleus and helps the vector to integrate into the host chromosome. Occasionally (up to about one-third of all eggs), one or more pole cells incorporate the transposon in the DNA and so the flies that emerge from the injection (the G0 generation) carry the transformed chromosome in some of their eggs or sperm. In the case of the *white*+ vector, *white*− hosts are used and the G0 flies always have white-coloured eyes. The next generation (G1) will contain a few red-eyed flies and these carry the gene of interest (Figure A1.9).

Now to the history: the prospect of transforming flies was an enticing one and scientists sought to find a way to do it as if they were searching for the golden fleece. There were various attempts that first looked promising but then did not fulfil expectations. One was by Allen Fox and Sei Yoon, who attempted to induce transformation by injecting whole wildtype DNA into mutant hosts. They used the *vermilion* gene and were excited to find flies that responded and had *vermilion*+ eyes. The results did not look nearly as good when they did the control experiment of injecting DNA from **mutant** donors and got just as many revertant flies! Eventually it was found that the DNA, a reasonably potent mutagen, was inducing mutations at a suppressor locus — and that these suppressed the *vermilion*− mutation.

There were two important parts of the discovery of the definitive method; first, the realisation that in flies there are transposing elements that can jump about; and second, the development of a practical technique to subvert them as agents to carry DNA into the chromosome.

Bill Engels and Mel Green, influenced by the famous work of Barbara McClintock on transposable elements in maize, had independently proposed

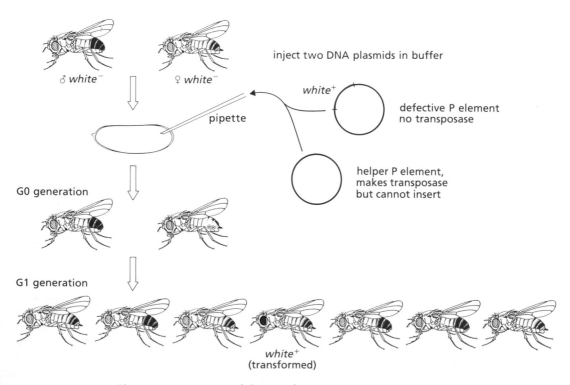

Figure A1.9 Summary of the transformation protocol using a plasmid carrying *white*+.

that transposable elements could be responsible in *Drosophila* for a phenomenon called 'hybrid dysgenesis' — it had been observed that a large number of mutations turned up when particular strains of wildtype flies were crossed. Engels and Green suggested this could be due to special bits of DNA that are released from the chromosomes of one strain, and subsequently reinsert elsewhere in the DNA, occasionally causing mutations.

In June 1980, at the main Cold Spring Harbor Symposium, New York, Rubin was waiting in line at the cafeteria with Engels and Margaret Kidwell and he suggested to them that if mutants produced by hybrid dysgenesis were cloned, one might actually find the elusive transposable element. Rubin and Kidwell collaborated with Paul Bingham and were able to purify bits of P elements from flies carrying suitable *white* mutations. Rubin realised, again stimulated by McClintock's results, that the P element proper should contain a gene for the transposase but the bits they had found were too small for that. So he looked for the complete P element by collecting appropriate pieces of DNA from the P strain; he found that a larger element of 2.9 kilobases was present in several copies.

Allan Spradling and Rubin used to talk a lot about science — good ideas often come out of rambling conversations between the right people. Rubin remembers: 'Given the known ability of bacterial transposons to carry antibiotic resistance genes around, as well as the trendiness of retroviral vectors for mammalian cells, it was fairly obvious, even at that time, to consider the use of P elements as vectors for gene transfer.' Why not simply inject purified P elements into embryos and hope they would hop about — but how would they detect the hopping at a reasonable rate? Fortunately, Engels had found an unusual mutation of the *singed* gene that was itself incredibly sensitive to hybrid dysgenesis. As many as 50% of the flies coming from a dysgenic cross showed a change at the *singed* locus; the mutation, *singed^w^*, therefore made a responsive assay. Spradling injected their P element into *singed^w^* flies and had to go through 6000 progeny before he found the first altered one — but he and Rubin showed both by analysis of the DNA and by *in situ* hybridisation to the chromosomes that the P element had inserted. They kept very quiet about their exciting results: 'We wanted to have evidence that would convince even the most sceptical critic: there had been false claims of *Drosophila* transformation in the past and we wanted to be sure.'

They also wanted the P element to carry another gene, and they chose the *rosy* gene. It had been cloned and it looked small; they were fortunate that the 8 kilobase piece of DNA contained everything necessary for function. They made a P element that contained that 8 kilobase piece. They then injected this into *rosy⁻* embryos and found *rosy⁺* flies in the G1 generation. 'When we saw that each of these transformants carried a *rosy* gene at a new site in the genome all doubts vanished and we announced our results in a departmental seminar in February 1982.'

As is abundantly illustrated in this book, transformed flies are very useful; one can see if the gene inserted has enough DNA to rescue mutant stocks, and thereby work out the size of the gene or indeed prove that the DNA you have really is the right piece (p. 16). One can make several extra doses of the gene to see if excess product affects the phenotype. One can mutate or delete parts of the sequence of the gene in both controlling or coding regions to find which bits matter. One can make hybrid genes where the promoter of one directs the expression of another (e.g. heat shock promoters, p. 56). The P element technique is one of the key methods in molecular genetics. Already most *Drosophila* workers make use of it and it will continue to be invaluable in the future. Not many methods are so fruitful and adaptable while being easy to apply.

List of quotations

1 Stern, C. (1954) Two or three bristles. *Am. Sci.* **42**: 213.
2 Chekhov, A. *The Three Sisters*, Masha, Act II. Translated by M. Frayn.
3 Carter, H. as quoted in Reeves, N. (1991) *Complete Tutankhamun*. Thames and Hudson, London.
4 Shakespeare, W. *Troilus and Cressida*, Act III. iii, lines 182–183.
5 Sander, K. (1976) *Adv. Insect Physiol.* **12**: 167.
6 Thompson, D.W. (1942) *Growth and Form*, p. 1036. Cambridge University Press, Cambridge.
7 Brenner, S. as quoted in Judson, H.F. (1979) *The Eighth Day of Creation*, p. 219. Jonathan Cape, London.
8 Maier, D., Preiss, A. and Powell, J.R. (1990) *EMBO J.* **9**: 3964.
9 Frankel, J. (1989) *Pattern Formation, Ciliate Studies and Models*. Oxford University Press, Oxford.
10 As quotation 6, p. 1085.
11 Seydoux, G. and Greenwald, I. (1989) *Cell* **57**: 1242.
12 Ready, D.F., Hanson, T.E. and Benzer, S. (1976) *Dev. Biol.* **53**: 217.
13 Carter H. as quoted in Reeves, N. (1991) *Complete Tutankhamun*. Thames and Hudson.
14 Shakespeare, W. *King Lear*. Act IV. vi, line 115. (The next two lines could be an encouragement for fly geneticists.)
15 Morgan, T.H. (1897) *Wilhelm Roux's Archives* **5**: 582.
16 Dryden, J. (1696) The epilogue to *The Husband his own Cuckold*, lines 35–37.
17 Eker, R. (1929) *Hereditas* **12**: 221.
18 Brasted, A. (1941) *Genetics* **26**: 347.
19 Tokunaga, C. (1961) *Genetics* **46**: 176.
20 Garcia-Bellido, A. and Santamaria, P. (1972) *Genetics* **72**: 87–104.
21 Bateson, W. (1894) *Materials for the Study of Variation Treated with Especial Regard to Discontinuity in the Origin of Species*, p. 85. Macmillan, London.
22 Struhl, G. (1984) *Nature* **308**: 455.

Other quotations given in the historical short stories come from letters to the author by the person quoted.

Index

Page numbers in *italics* refer to figures.

Hybrid dysgenesis 221
Hybridisation *in situ* 16, 30, 52

Imaginal discs 20
Induction 126, *128*
int-1 18, 102
Integrins 196
Intercalation 142, 150
Intragenic recombination 83−4
invected 210

Keilin's organ 20, 172, 214
knirps 57, *58, 59*, 60−1, 64, 68
Krüppel 17, 33, 45, 57, *58, 59*, 60−1, 68,
    70−1, *72−3*, 74−5, 129, 133, 172

labial 126, *128*
lacZ 32, 52−3, *54*, 70, *71*, 96−7, 163
Landmark bristles 162−4
Lateral inhibition 162, 167
    role of *lin-12* and *Notch* 170−2
lethal (1) polehole 41, 64
lin-12 170−2
Locust 100, 117, 118

Marking methods, cells 78−80
Maternal effect and rescue 1, 3, 8, 38,
    202−3
Mcp 113, 124, *125*
Mesoderm 13
    segmentation 93−5
    somatic *see* Somatic mesoderm
    visceral *see* Visceral mesoderm
Minute 80, *81*, 82, *83*, 86, *87*, 89, 93, 123,
    136, 148, *149*, 208
Mitotic recombination 82−6
    cell markers 86−7
    eye 188, 190−1
    mesoderm 93
Molecular epistasis 45, 57
    ftz expression 97, 99
Morphogens 27, 57, 63, 67, 69, 138, 140,
    201, 203
    history 206
Mosaics, genetic 10
    cell markers 86−7
    mitotic recombination 82−6
    nuclear transplantation 123
    pole cell transplantation 35−6
multiple wing hairs 82, *84*, 86, 209
Muscles
    pattern, determination 122−6
    source 21−2
    thorax, origin 93−5
Mutagenesis 131−2
Mystery cells 186

naked 102−3

nanos 25, *34*, 35, *37*, 40−1, *42*, 50, 57−8,
    59, 60−1, 68
Neurectoderm 119−21
Neuroblasts 172−7
Neurogenesis and spacing patterns
    158−9
Notch 5, 6, 7, 85−6, 176, 187, 196
    lateral inhibition 170−2
    spacing patterns 166−8, *169−70*
Nuclear transplantation 123
nudel 43, 44
numb 176
Nüsslein-Volhard, C. 42, 50

Octhera mantis 152
odd-skipped 100
Ommatidium 180−2
    development 185−7
Oncopeltus 91, 138, 140, *141, 147−8,*
    158−9, 160, 173, 176, 187
oskar 9, 32, 34, 38, 68

P element 219
Pair rule genes 69
paired 69, *99*, 100
Parasegments 4, 18, 19−20
    cell allocation 91−2
    and gradient fields 147−8
    history 214−5
    in the mesoderm 121
Parydra aquila 152
patched 33, 102−3
pawn 86−7, 123
pelle 43
Peripodial membrane 21
Phenotypic suppression 114
Photoreceptor cells 180−1
pipe 43, 44
Polar plasm 9
Pole cell transplantation 35−6
Pole cells 9−13
Polyclone 80
Polycomb 107
porcupine 101
Positional information 146, 147−8
postbithorax *108, 116,* 212
Posterior system 33−8
Proneural cluster 162
Proportion, genetic control 152−6
pumilio 38

raf 64
Reporter genes 52
Rhodnius, experiments 138, *139,* 140,
    145, 159, 206
Rhombomeres 218
RNA, *bicoid*, localisation 30−3
rosy 221
rough 191, 218
runt 100